W9-ARC-398

Gene Logsdon

Successful Berry Growing

How to Plant, Prune, Pick and Preserve Bush and Vine Fruits

Rodale Press, Inc. Book Division
Emmaus, Pa.

Second Printing—October 1974
OB-421
Printed in the United States of America on recycled paper.

Library of Congress Cataloging in Publication Data

Logsdon, Gene.
Successful berry growing.

1. Berries. 2. Viticulture. I. Title.
SB381.L63 634'.7'0973 74-10635
ISBN 0-87857-089-6

Dedicated to Wendell Berry
whose poetry and essays
have eased the soreness
from my mind.

I am not bound for any public place,
but for ground of my own
where I have planted vines and orchard trees,
and in the heat of the day climbed up
into the healing shadow of the woods.
Better than any argument is to rise at dawn
and pick dew-wet berries in a cup.

from "A Standing Ground,"
a poem by Wendell Berry
from his book *Farming: A Hand Book*,
Harcourt Brace Jovanovich, Inc., New York

Contents

The strawberry—firm, ripe and tasty—aptly symbolizes the wonderful world of berries. The satisfaction that comes from growing your own berries in your own organic garden is not difficult to achieve.

Chapter One

The Wonderful World of Berries

Growing berries ranks as the most rewarding of all gardening and farming experiences. Let me count the ways—as a poet has said it. The sight of a blueberry bush, laden with misty blue fruit, is certainly as pleasant a sight as a rose arbor in bloom. Black raspberries in July sunlight sparkle like jewels. And what flowering shrub can match the creamy spray of elderberry blossoms?

Even when berry plants aren't bearing, they make fine ornamentals, especially blueberries and whortleberries with their red fall leaves. Strawberries, particularly evenbearing varieties which do not runner much, make good border plants in flower gardens. In the north, currants and gooseberries provide early leaf and bloom.

If berries reward the eye, then what the homegrown variety does for your palate is beyond description. Rarely can you buy berries that taste as good as the ones you raise. Even with strawberries, which the commercial growers now do a remarkable job of marketing the whole year round, your own will taste better most of the time.

Some of the finest berry eating, such as red raspberries in

the fall, can rarely be purchased, and then only for a king's ransom. Not even kings can enjoy the yellow raspberry (the best-tasting of all fruit in the world in my opinion), unless, like you and me, they plant some in their gardens. I've never seen them sold except in Lancaster County, Pa.

Profit From Small Fruits

Berries are financially rewarding too, whether you raise them only for your own family or for sale for sideline income. My family managed to consume 100 quarts of strawberries this year that would have cost us at local prices in the supermarket a minimum of $75. Our cost: $14 for plants and a little labor. My son and I have just spent 15 minutes picking 4 pints of blueberries from our patch. Store price on a pint of blueberries is 70¢, my wife tells me. So, our 15 minutes were worth $2.80, or $11.20 an hour—pretty good wages, especially since we spend very little time on the blueberries other than picking them.

The small farmer, the modern homesteader from the city or anybody else looking for more income from his land can try no more practical crop than berries. From an acre of berries—any popular eating berry—sold retail at a roadside stand, a grower should be able to gross $2,500 with just a little luck and knowhow. And he can accomplish this on an acre without having to buy the expensive machinery or facilities that put so many other types of farming operation beyond the small farmer's pocketbook.

To give you some idea of what the far out possibilities of berry profits could be, take a look at the yields that California strawberry growers are hitting these days. It may seem unbelievable, but harvests in that state sometimes reach 20 *tons* of berries per acre. If you had 1 acre of that kind of crop, and 4 or 5 family members to help you harvest it, and could sell the whole production at 75¢ a quart (price of strawberries around here), your gross profit would be an amazing figure in excess of $10,000. (A quart of strawberries weighs more than a pound, but less than 2 pounds—depending on how you fill the quart basket.) Of course strawberry growers incur colossal overhead in order to hit yields like that. Large operations would consider themselves successful if *net* profits from strawberries move above $800 per acre. Small growers with only 1 or 2 acres and their own labor can do better than that. How much better? The answer to that is the number of tons of berries per acre *you* can raise and sell.

Growing raspberries can be extremely lucrative too, if you have a ready market. The berry will not keep very well, even when refrigerated, and it is usually imperative to sell it the same day it is picked. A fellow I've known for a number of years (now retired) worked a 14-acre berry farm. On the Great Plains farmers need nearly that much land just to turn their big grain rigs around on. But my friend concentrated on raspberries on his little patch of ground and did so well he usually spent the better part of the winter vacationing in Florida. Oh, he never handled huge sums of money the way a big farmer or cattleman might, but most of what he handled, he kept. He always maintained that when he sold his berries retail, he was able to net $1,500 per acre on the average.

Fruits of the Seasons

Another nice thing about berries is their continuity through the entire growing season. Here in eastern Pennsylvania, our berry season begins in the last week of May. We bicycle down the road past Driscoll's barn, where the barnswallows always flit out to meet us, and pick a few handfuls of wild strawberries on the bank above the road. About the first of June, the first big strawberries ripen in our garden and we battle the brown thrasher family for the privilege of eating them.

Carol Logsdon, the author's wife, gives the family strawberry patch a going over. The harvest is eaten in cream, on shortcake, as preserves or as is.

By the last week in June, the strawberries are slacking off and the black raspberries are coming on strong—big ones in our garden, smaller ones in the woods. No sooner have we picked the first flush of the blackcaps than the red raspberries and yellow raspberries begin to ripen too.

By then the middle of July has come and passed. The mulberries have been dropping from the trees for a week or two, making a mess on the road that runs in front of our house. But we don't mind, because a heavy mulberry yield means the birds will not eat so many blueberries, which we are now picking as fast as they ripen. At the same time, wild wineberries are ready to pick along Gypsy Hill Road or in most any brush land along Wissahickon Creek. And if blueberries and wineberries don't grow in your part of the country, you can be picking currants and gooseberries instead.

The last days of July and the first part of August is blackberry time. We have big ones in the garden and all we could possibly pick growing wild in the many abandoned fields of the neighborhood. Along Meetinghouse Road, you can pick blackberries without getting out of your car—if you're that lazy!

In August too, the elderberries ripen. I must have an elderberry pie at least once a year. Out of the remaining berries we make jelly—to my taste, the best berry jelly of them all.

Then in September, the fall (sometimes called everbearing) red raspberries begin to fruit. Properly handled, they will go on bearing until frost.

Cranberries ripen in the fall too, and although not generally grown in gardens, they *can* be grown anywhere blueberries grow. I don't like them enough to go to the trouble, but if I lived anywhere close to a wild cranberry bog, I'd certainly avail myself. As it is, I'm content to search for the lingonberry, which is the cranberry's little wild cousin. It doesn't need a swampy locale or a bog; I've found them on hillsides in Maine.

At any rate, we have a full 5-month season of fresh berries. Certainly farther south and on the West Coast you could do even better than that. While the long berry season sounds like work, the mere fact that the season does stretch out enables you to level the work load. In fact, if you do a little berry work every month of the year, like you should, the job won't seem bad at all.

Gardening Seasons

January is planning and ordering time. In February I begin pruning the bramble berries and finish that work in March. In April I plant new strawberries. In May I rogue out sucker canes from the red and yellow raspberry rows and do any raspberry or blackberry transplanting that needs to be done. June is busy with strawberry picking, plus pruning and blossom-pinching the new plants that will produce next year. July is also mostly a harvest month, but toward the end of the month you must cut out old raspberry canes and burn them. August is the time for finishing that job, as well as pinching back next year's bearing black raspberry and blackberry canes. In September the fall red raspberry season starts; it lasts through October—or until frost. In early fall you can train next year's bearing blackberry canes to trellises or stakes, if you wish. November is mulching month on my calendar—the time I add a fresh supply of leaves and/or manure to blueberries, gooseberries and all the brambles. Also in November I cut down all the fall red raspberry canes that have borne fruit, even though some of them would bear again next summer. (I'll tell you why I follow this practice in a later chapter.) Come Thanksgiving time, we sauce the last berries of the season, cranberries. I haven't tried to raise cranberries —if we want our own, we get them on our annual fall trip through the Pine Barrens in New Jersey. Or we pick wild lingonberries if we do fall hiking where they grow. In December I cover strawberries with straw and finish up any mulching remaining to be done on other fruit. I also trim elderberries in December, though that job could wait. Because my boysenberries would freeze in our winters, I gather the canes together in December, lay them flat on the ground and cover them thickly with mulch.

Then the snows fly. We haul out the nursery catalogs and prepare for another cycle of seasons with our favorite fruits.

Sounds like a lot of work, but if you do a little every week, you'll find that you can take each task in stride and enjoy it. The good gardener knows that a half-hour every day will handle a surprising amount of gardening—just as the writer who writes just 3 pages every day can complete quite a stack of books in a lifetime. The Un-gardener doesn't grasp the old saying: "Slow and steady goes far into the day." He becomes excited at the prospect of raising "most of the food I need in the backyard." He orders enough seed to choke a horse, and for about 2 weeks in May or even maybe 3, he is a whirlwind of furious activity. He spades, he hoes, he sweats, he plants, he weeds. If all goes well, he has a nice looking garden by June. But other summer fun now absorbs his attention: a swim today, over to Joe's tomorrow evening for a cookout, can't miss that film the next night, golf on Saturday, 3-day business trip next week and so forth. Instead of reserving that half–hour or 45 minutes every day, the Un-gardener puts off the work until he is faced with a growth of weeds that demands a whole weekend's time. Six straight hours of cutting weeds already too large for the cultivator to handle lessens the Un-gardenerer's enthusiasm considerably. It'll probably be another week before he again attacks the weeds and bugs. Then, just as he is about to start, a big rain falls, preventing garden work for another three days. By that time, the weeds are too far ahead for him to keep up. By August weeds and grasses have carpeted his "garden" and bugs have eaten everything else. His box score: smartweed–700; red-root pigweed–4,387; strawberries–14; tomatoes–5; errors–too numerous to mention.

The short time you spend every day on your berries is good for you. Bending, stooping, squatting, flexing muscles—all are exercise as valuable as yoga. And you can get tanned in more places than you can at the beach. And keep your weight down and your body fit without jogging along the road where the automotive monsters can run over you. When was the last time you saw a good gardener who was soft and flabby?

Organic Growing Techniques

Berries should be the special favorites of all *organic* gardeners and farmers, for the simple reason that you can raise them with organic methods a whole lot easier than tree fruits. I have never had to spray insecticides on my berries—any of them—in the ten years that I have grown them seriously. I can say that—not necessarily out of any organic conviction—but simply because there weren't enough harmful bugs to warrant using an insecticide. And that certainly has not been true of my tree fruits. Without at least dormant oil spraying of fruit trees, the organic orchardist would get very little fruit.

Because gray rot, a fungal disease, is especially bad on my strawberries in wet years, I've tested chemical fungicides on them. I've sprayed half a row and left the other half unsprayed. Result? You couldn't tell the difference. If the weather remained wet, the rot got worse; if hot, dry, sunny days followed, the rot diminished—on both the sprayed and unsprayed portions.

Getting along without herbicides is the least of the organicist's worries. It is more work to cultivate or mulch for weed control than to spray, but the rewards to your soil are greater too. Moreover, weeds are not the same kind of threat to you as are plant diseases and insects. With weeds you are never helpless—you can always control them, given enough hours in the day. But there are plant diseases you can't control, even with chemicals.

There's proof from nature that berries can produce without chemical help from man. The maggot problem in wild blueberries in an exception, but in general, despite all kinds of wilt, viral disease, bugs and low fertility, the wild berries go right on pluckily making a crop almost every year. And you will often find some of the best wild berries on poorer ground.

This last observation has some significance for organic growers. Domesticated berries need fertilization, I'll be the first to say. Yet my experience is that most berries will thrive without using the high–powered commercial fertilizers that most commercial farmers consider an absolute necessity. Since I don't raise berries *for a living,* I'm not about to argue with those who do. But I do feel sure that if you wish to grow berries organically, you can expect success using manures and mulches for fertilizer.

In my experience I have found it easy to over-fertilize berries. Too much nitrogen will cause strawberry plants to grow very lush, but not necessarily to produce more berries. And I *think* that strawberries given high feedings of nitrogen tend to rot quicker in wet weather. Raspberry canes—especially black raspberry canes—that grow exuberantly in the late summer from an overdose of nitrogen will often freeze out during the winter.

Mulching and Soil Fertility

If you are a mulch gardener already, you can apply that knowledge and experience to excellent effect on your berries. Berries love mulching—love the protection it gives from winter freezing and thawing, the moisture it retains in dry weather, and above all, the organic matter it adds to the soil over the years. I keep my berries mulched year around. It keeps weeds under control at all times and saves me hours of cultivation. No berry will endure long stretches of dry weather, but mulch will carry them over short periods of drouth in fine shape, especially where mulch has been used for 5 years or more and therefore has raised the organic matter content in the soil below it. I believe that in 5 years mulching becomes a substitute for about 4 inches of irrigation water during a time of critically dry weather. In other words, with continuous mulching, the commercial grower can profitably raise berries east of the Mississippi 8 years out of 10 without supplemental irrigation, and the small gardener can do it 10 years out of 10. Mulch is far cheaper than an irrigation system. Moreover, after 5 years of mulching, the continuing low-nitrogen charge of fertilizer the mulch adds to the soil is *just about all the fertilizer your berries need,* especially if your mulches are, or contain, animal manures.

Any mulch that you have found beneficial in your vegetable garden will be fine for your berries. My choices are very much influenced by what I can get for the least cost.

On strawberries I favor horse manure: strawy horse manure the first year, and clean straw the second or bearing year. White-pine needles are good too. On blueberries I use sawdust mixed with manure or rotted sawdust alone, if I can get a load of it free from the nearby sawmill. I mix the manure with fresh sawdust because fresh sawdust will rob nitrogen from the soil until it rots unless you provide supplemental nitrogen. (But on ground that you have mulched for years, there's enough fertility to handle the rotting process without supplemental nitrogen.) Oak leaves make good blueberry mulch too—oak leaves are slightly more acid than most other tree leaves and blueberries like acid soil. On brambles (raspberry, blackberry, boysenberry) I mulch with any leaves I can gather from around the neighborhood. Grass clippings are fine, but rot away fairly fast, so I use them on the vegetable garden.

Old rugs and newspapers can be utilized as mulch around permanent berry plantings; if you use them, cover them with leaves or another natural-looking mulch so your garden doesn't look like the city dump. Rotted hay, old hay bales and salt hay are fine mulches, though the two first may contain weed seeds. Shredded corn stalks, sorghum stalks and sugar cane stalks make fine mulches. Sorghum-sudan hybrid grasses that make such fantastic growth and can be cut 4 or 5 times a year for hay can also be cut for mulch—an idea you might think about if you are short of outside sources of mulch.

Though hard to get except in Kentucky and North Carolina, tobacco stems are an excellent mulch since they are

Raspberries are an excellent small fruit for the home garden, as the fragile flavor and texture is too easily lost in any but the most cautious handling. Here, the author's son, Jerry, plucks a red raspberry.

high (compared to other organic sources) in potassium and nitrogen and are also a slight deterrent to some insects because of their nicotine content.

While you are applying the mulch, you should have some idea of its value as fertilizer. This is not always a matter of simply referring to a list in a book on soil fertility. For example, the amount of nitrogen in grass clippings will depend quite a bit on the state of the soil the grass grew in, and on the amount of moisture available as it grew. If the lawn is heavily fertilized, the grass clippings will contain more nutrients. The third cutting of alfalfa for hay in late summer usually has a higher protein content than the first cutting because first cuttings grow faster and have a higher water content. However, nutrient analyses, such as the figures I use, (from Rodale Press book, *Organic Fertilizers: Which Ones and How To Use Them*), are valuable as a way to *compare* one mulch with another, giving you some idea of what you might need to attain *balanced* fertility. Balance is the key.

Cottonseed meal contains 7% nitrogen (N), 2½% phosphorous (P) and 1½% potassium (K), the 3 main elements in plant nutrition. That makes cottonseed meal a good source of nitrogen. Fish scraps contain about the same amount of N, but up to 13% P, very high for an organic fertilizer, and 3% K. Tobacco stems contain 3% N, 1% P and a high 7% K. Seaweed contains almost 5% K. So a combination of cottonseed meal and fish scraps with tobacco stems or seaweed would make an excellent, balanced fertilizer. It would also be expensive to purchase, though gardeners in states along the ocean could get both the fish scraps and seaweed free. (Those Indians who buried a fish in each hill of corn they planted knew what they were doing.)

Bloodmeal, bonemeal and tankage (meat scraps) are all good sources of N and available almost everywhere. But they are expensive to use in quantity.

Animal manures, if you can haul them yourself and use plenty, are the most economic mulch fertilizers. Rabbit manure is the richest—which means it contains the most nitrogen: 2.5% N, 1.5% P, .5% K. Poultry manure runs 1.1% N, .8% P, .5% K. Sheep and horse manure analyze about the same: .7% N, .3% P, .9% K. Duck manure: .6% N, 1.4%P, .5% K. And cow and pig manure about the same: .6% N, .2% P, .5% K.

Animal manures are low in nutrients compared to inorganic commercial fertilizers, but numbers don't tell the whole story. If you like to experiment, try this one to prove something to yourself. Take two equal plots having the same kind of soil. Fertilize one with 10–10–10 commercial fertilizer (or any analysis that is right for that soil) at the rate of 200 pounds per acre; on the other plot apply animal manure at the rate of 10 tons per acre. Plot A, with commercial fertilizer, will have—according to the numbers—a good deal more plant nutrients than plot B. And crop yields will be better on plot A than plot B, all other things being equal. Now, the next year put *no* fertilizer of any kind on either plot. All other things being equal, yields from the two plots will be about equal. But in the third year, with no fertilizer whatsover applied to either plot, the manured plot will have a better crop on it. What that proves, I think, is that manures have "staying power"—they energize the soil to fertility beyond their analyses. Manures also maintain proper amounts of trace elements in the soil. The grower relying on inorganic fertilizers must often add trace elements to his NPK treatment.

Things sometimes happen in my garden as a result of mulch-fertilizing that are hard to explain. My soil is not by far the best in the East. When I came here, it would produce only poorly. Two years ago I covered with a foot of leaves a plot of ground that had not been farmed for 100 years, but that had in fact grown back to brush and been cleared again. The ground may never have been fertilized with anything. Anyhow I was too lazy to plow up that tough sod, so I just covered it with fall leaves. The next spring I put on about a 5-inch layer of manure and set some tomato plants down through the mulch, disturbing the soil only enough to set the plants. The crop was normal. But this year I tilled the plot with a rotary tiller (the sod having all rotted away), planted tomatoes again and again mulched with a good 6 inches of manure.

Those tomato plants are now rather frightening. Today is July 25. The plants stand, *without any staking or trellising whatsoever, over 4 feet tall.* And they're still growing upward. To support that kind of height, each plant is spread out over an 8-foot-square area—and still spreading. I have not pruned even one sucker. The plants are loaded with fruit.

I have raised tomatoes on good land; I've raised them by socking inorganic fertilizers into both good and poor land; I've raised them by gosh and by luck. But never in my life have I seen such results. And hardly 100 feet away, my neighbor's tomatoes, grown in more conventional ways, are spindly and miserable. (Forgive me, Ed: I gloat only for the sake of scientific progress.)

I want the experts, especially the commercial fertilizer experts, to tell me why my tomatoes are growing like they are.

Among the various kinds of leaves that may find their way into your mulch, raspberry leaves have an unusually high (for leaves) nitrogen content—1.35%. That's why, when my raspberry plants are healthy, I will sometimes let the old canes die and drop their leaves before I remove them. That's also why (again if the plants are all healthy and insect-free) I will shred the fall-bearing red raspberry plants after frost and let the shreddings lie right there on the ground where they can do some good.

Oak leaves run about .8% N, .35% P and .15%K. Apple leaves, 1%N, .15% P, .35% K, are a bit better than pear leaves at .7% N, .12% P, .4% K. I have not been able to find out what the nutrient value of a strawberry plant is when worked into the earth as green manure after the berry season, but I'm inclined to think it has some significant value. My late corn doublecropped after strawberries always does unusually well.

So much for the basics behind the labor and the soil management of berry growing. Now let's get down to the culture of specific kinds of berries.

Chapter Two

The Strawberry: A Better Berry There Never Was

I can think of 2 very good reasons why every gardener or farmer should grow strawberries. The first reason is a bowl of strawberries and cream; the second is strawberries, cream and shortcake. The berries should be picked fresh and ripe in the cool of the day, the cream skimmed from a bucket of Jersey milk that has spent the night in a cooler or springhouse and the shortcake should be the kind my Kentucky mother-in-law makes, short—thin, if you will—crunchy, nutty and still warm from the oven. In this context the strawberry does not have to have any more excuse for existing. Nor does the gardener who grows it. He follows the highest calling.

Fortunately for mankind, the strawberry is a very adaptable fruit; there are varieties that will perform well for every region of the United States. No other fruit is so widely distributed, except perhaps the apple.

But the strawberry is not only at home in nearly every state. Its versatility extends to nearly every kind of gardening. It can be grown in pots, almost like a house plant; it can function as a border plant in a flower garden; it will produce well in beds or in garden rows; it can be raised commercially in large fields; it grows wild. In Japan the strawberry is even cultured profitably in greenhouses.

Moreover, the strawberry comes into production quicker than any other fruit I can think of. The year after you set out plants they hit full production. More than one beginning fruit farmer has lived on the money he made from strawberries while getting his orchards established.

Most strawberry varieties like cool weather, but will put up with hot. All strawberries like plenty of moisture, but will endure a little drouth if they have to. Strawberries prefer a slightly acid soil with a pH of around 6.0, but they will tolerate a fairly wide range of slightly acid–slightly alkaline soils —from a pH of around 5.5 to 7.5.

Some growers say strawberries prefer lighter soils, but from my own experience I know they will do quite well on heavy, clay soils too, if such soils have *a high content of organic matter.* The best wild strawberries I've ever found were on poor, worn-out soils of abandoned farms. Tame berries will respond to a *balanced* fertility program, but do not seem to need expensive high-nitrogen fertilizers. A strawberry plant will eat nitrogen the way a child eats candy, and the results can be equally as poor.

In short, if you give the versatile strawberry the proper tools—water and organic matter—your shortcake should be properly decorated no matter where you live.

I'm about to try to tell you just how to grow strawberries, but I've got a problem. I am sitting before my Master, the typewriter, staring out the window at my wren house. I dutifully nailed the nesting box together this spring just like the bird book said to. No wren has darkened its door yet, but a dapper cardinal is eyeballing it for size. Or maybe he is just wondering what kind of creature would build such a structure. The house, though, is not my problem—nor is the cardinal except symbolically. I am only half aware of both (though

Strawberries are one of America's favorite fruits. Though few people eat them for it, strawberries have high nutritive value. Half a cupful of strawberries contains at least as much vitamin C as an average orange.

having half my mind on that does not leave very much for the problem) as I try to figure out how best to *begin.* No matter where I start, I find myself engaging in a lot of putting-the-cart-before-the-horse. So, I think I will assume that you have not grown strawberries before (just as I had not built a wren house before) and will show you, step by step, how I do it. If I tell you something wrong, it will at least be my own mistake —not one copied out of another book. But I'm not too concerned about my possible errors. We always have plenty of strawberries every June. And you will too.

Selecting the Right Variety

Pretend it's January. That's when your adventure with strawberries should begin, because that is when you order your plants. By ordering early you assure yourself of getting the varieties you really want. (In 1973, those who ordered late learned that even shortages of strawberry plants can develop.)

What varieties to order? That depends on a number of

factors: 1). The quality, taste and physical characteristic of the berry itself; 2). the area of adaptability—the right variety for the right locality; 3). blossoming and ripening time—early, midseason and late.

Whichever of these 3 factors is most important depends somewhat on your purpose and your personality. Nonetheless, you must use all three of them to make the right decision in picking your selections.

. . . for Flavor and Storage

A berry that tastes good (quality factor) is my first consideration—the hedonistic impulse in my blood is to blame. And of course taste is, well, a matter of taste. Over the years, I've tried nearly every variety of berry I could buy and often found that the varieties that tasted best to me often had some other characteristic I didn't like. In other words, though every strawberry approaches perfection, none quite makes it all the way. I include Fairfax in all my plantings because I like its taste best—even though it is not the most disease resistant berry I can grow, nor will it keep very well after it is picked, nor will it freeze very well anytime.

Keeping ability, shipping ability, freezing ability—these are all parts of the quality factor which may be important to you. Berries that keep well are usually firmer with tougher skin; commercial growers like them because they can be shipped half way around the world without apparent harm. For your own table or for your own close-at-hand market, these kinds of berries are *not* your best choice.

Good varieties for freezing? I can't tell you because we have never found one that we like frozen. So we freeze other kinds of berries and eat our strawberries fresh or in jams. Some nursery catalogs recommend the following varieties as good for freezing: Midland, Pocahontas, Redglow, Sparkle, Surecrop, Earlidawn, Fletcher, Redchief, Earlibelle and Northwest. Not the best, but "above average" for freezing, says the Department of Agriculture, are Catskill, Hood, Marshall, Midway, Puget Beauty, Siletz and Tennessee Beauty.

When you start asking strawberry-men which varieties are best for preserves, they won't always agree. I go to the opposite extreme in this case and think that most any variety makes good preserves—and that because of the way most homemakers drown their preserves in sugar, you can't tell any difference between varieties when jammed anyway. Among people with more discernment than I have, Blakemore, Earlidawn, Pocahontas, Sunrise, Tennessee Beauty, Hood and Marshall are supposed to make better jams. Throw in Albritton, Guardian and Redchief for good measure. If you have the will power not to eat them as you pick them, wild strawberries make delicious jams and jellies too.

For just plain good eating, I've mentioned Fairfax. I like Sparkle and Catskill too, but I'd hate to have to do much ranking beyond that. You have to be careful naming names (just like in politics) of flavorable strawberries because a particular variety may well taste better when grown in one area than when grown in another. For instance, my favorites, Fairfax and Sparkle, are good dessert varieties in the North, but not always in the South. Midway has more flavor in New York than in Maryland. Marshall tastes better in Oregon than in lower California.

. . . *for Your Locality*

Just as commercial growers have their favorite varieties in every part of the country, so do hobby gardeners. In the South, more gardeners grow Blakemore than any other variety, according to Department of Agriculture statistics. In Virginia, West Virginia, Kentucky and Tennessee, it's Tennessee Beauty. In the Carolinas, eastern Virginia, northern Georgia, northern Alabama and northern Mississippi:

Albritton. Headliner and Florida Ninety, the commercial berries of the Gulf Coast, are also popular in home gardens along the coast and in Florida. In and around Maryland and southern Pennsylvania, Fairfax and Surecrop are gardening favorites. In New England, New York, Ohio, Michigan and Wisconsin, gardeners rely chiefly on Catskill and Sparkle.

On the Great Plains, Dunlap, an old variety, is still popular. In the West, Marshall is the favorite home garden variety.

Aware of how a variety might taste different when grown in different climates, the second factor in making your selection—area of adaptation—becomes quite important. Here's a list of most available varieties according to region they grow best in. (Don't hold this guide infallible. It's a general listing, and just as sure as sunshine in July, someone has grown a variety successfully where it ain't supposed to grow successfully. And regions often overlap.)

Northern varieties: Midway, Catskill, Surecrop, Sparkle, Premier, Fairfax, Fletcher, Midland, Empire, Dunlap, Trumpeter, Ogallala.

Southern varieties: Florida Ninety, Dabreak, Headliner, Albritton, Earlibelle, Sequoia, Pocahontas.

Western varieties: Northwest, Shasta, Tioga, Fresno, Marshal, Siletz.

Eastern varieties: Guardian, Jerseybelle, Raritan.

The general rule is that a western berry won't do well in the East and vice versa, with the same rule holding true for southern and northern varieties. But within a region, one variety may do better in one place than in another. For instance, Sunrise and Surecrop usually grow well in the central

East, but not as well in the Northeast. Midway, on the contrary, produces better in New York than in Maryland. Florida Ninety makes big berries in Florida, smaller berries in North Carolina. In the plains states, like North and South Dakota, where people (and plants) have to learn to live with 20 below zero weather, Dunlap and Trumpeter are grown because they survive that kind of winter. Ogallala is a tough variety too. It has a reputation of surviving temperatures of a minus 40 degrees.

Ogallala is an everbearing variety. You won't catch me saying many nice things about everbearing strawberries because raising them has always been a disappointment for me. However, everbearers seem to do better in the North than in the South. This has something to do with summer day length: for instance, June 21, the longest day of the year, is nearly 3 hours longer in northern Minnesota than in southern Florida. The longer day lengths help the late summer crop of everbearing strawberries.

Some gardeners claim good luck growing the everbearing Ozark Beauty in the central states. Geneva is another everbearer which seems to have shown promise in New York. Here in Pennsylvania, Geneva made delicious berries for me, but they were extremely soft, prone to rot and did not yield well.

I've concluded that I was not a good grower of everbearing varieties and no longer bother with them. There are berries other than strawberries more satisfying to grow for late summer and fall production.

. . . for Your Growing Season

We've discussed 2 factors in strawberry selection, taste quality and area of adaptation. The third factor you should consider is the time it takes for a variety to ripen. You want as long a strawberry season as you can get, so among the varieties you grow should be early, midseason and late berries. In this regard, here's how the better known varieties are usually grouped:

Early: Earlidawn, Midland, Atlas, Blakemore, Dabreak, Dunlap, Sequoia, Sunrise, Earlibelle, Florida Ninety, Headliner, Premier, Redchief, Redglow, Surecrop.

Midseason: Aliso, Apollo, Catskill, Empire, Fairfax, Raritan, Shasta, Fletcher, Fresno, Guardian, Marshall, Midway, Tioga.

Late Midseason: Albritton, Armore, Hood, Northwest, Puget Sound, Robinson.

Late: Badgerbelle, Columbia, Garnet, Jerseybelle, Marlate, Siletz, Sparkle, Tennessee Beauty.

Very Late: Redstar, Vesper.

Weather and soil can often blur fine distinctions between ripening dates. I've seen Catskill listed as an early variety, Sparkle as a midseason variety (it is for me), and Surecrop as a midseason berry. A lot depends upon the weather and the way you handle your plants. If you have a late spring, early berries and midseason berries may ripen at about the same time. Berries will ripen more slowly in cool weather; quicker if you have a hotter than usual spring.

But by growing varieties with different ripening dates, you safeguard against getting your whole crop killed by a late frost. All the berries won't be blooming at the same time so a frost will only harm part of your crop.

To further safeguard against frost, I keep the mulch on my strawberries as late in spring as I can without smothering

Picking wild strawberries takes time and effort, but Jenny, the author's daughter, knows their taste makes the effort worthwhile.

new growth. This practice holds all the varieties back from blossoming time a few days—the better to avoid that last frost that might come. If I were selling berries commercially I wouldn't do that—I'd want some berries ripening as early as possible, when market price is usually higher.

Some varieties are less susceptible to killing freezes. Premier (sometimes called Howard 17) has a reputation for having the toughest blossoms—a sort of Jimmy Cagney of the strawberry blossom world. So long as the flowers are not all the way open, it, and Earlidawn, will resist a freeze that will kill other berry blossoms. Catskill and Midway blossoms sometimes escape frost because their flower stems are short. Their blossoms sometimes remain protected under the strawberry leaves where frost doesn't touch them.

Ordering Your Plants

You have now finished your brainwork on selecting varieties. You have narrowed the choice to three or four that are recommended for your area, have good dessert quality and varying ripening times. Now you have only one more decision to make. How many plants do you need? Or, how many berries do you want to produce?

For practical purposes, figure that each plant will produce a quart of berries. A good gardener may do much better than that, but that formula is a safe one. A family of 2 adults and 2 children, if they like strawberries the way we do, can consume a quart or 2 of berries every day during the season. You have to be the judge on how many more you want for jams, for freezing, for giving away or for selling.

At this point, you are ready to make out your order. After having gone through this thinking and planning process for 10 years, I will order 25 Fairfax plants (my favorite), 25 Jerseybelle or Marlate for a late variety and 25 Catskill or Sparkle, depending upon my mood at the moment for a midseason variety. I order in multiples of 25 because that's the way the nurseries offer them. The 75 plants give us about 100 quarts of berries, the amount we want. And these varieties give us the full season spread. I may order a fourth variety, or substitute for one of my regulars, to try out a new variety or experiment with an old one I haven't grown before. You should do that too—and will have to do it at least until you find the early, midseason and late varieties that grow the best in your particular soil and climate.

When you order, stipulate that you do not want the nursery to substitute other varieties than the ones you have selected (unless you really don't care). Nurseries usually give you this alternative in case they are out of a particular variety, or if that variety is in short supply. Since your order will be received in January, you will generally have no problem. The early bird gets the worm.

Be sure to include the approximate date you'd like to receive your plants—as close to the time you plan to plant them as possible. In the northern half of the country, that's usually between April 15 and May 1st, though you can plant later.

Nurseries have always been very good about sending plants on the dates I've requested. If I say April 18, that's when the plants usually arrive—never more than a day early or late.

If your plants arrive during a 3 day downpour—which is likely to happen—don't despair. Just put them in the refrigerator until the soil is dry enough to work. Don't open the package; just store them as they came. You can keep plants a month that way, but plant them as soon as you can to be on the safe side.

If you've ordered your plants to arrive on a weekend, try to make sure they don't sit in the warm post office over Sunday. We have a small post office (thank goodness for *small* things) and I always tell our postmaster I'm expecting plants on Friday. If they are late and don't arrive until after I've made my Saturday morning visit to the post office, he is kind enough to give me a call so I can get them before he closes in the afternoon.

Groundwork

Preparation of the soil for planting should be started in the previous fall. Fall plowing or rotary-tilling allows the ground to dry out faster in the spring. Gardeners should have plowed under at that time a green manure crop of clover or rye grass and any animal manures available. An application of rock phosphate can be applied at that time too. If the soil is very acid (below a pH of 6), sprinkle agricultural limestone on top of the worked ground. Don't plow lime under, as it will leach down fast enough anyway.

Now, in the spring, rotary till or disk the ground again. Don't get in a hurry. If the earth makes mud balls when you try to work it, the ground is still too wet.

Before you set your plants, you decide what spacing between plants and between rows you want to maintain. There is no single right way to do this, but there are several *good* ways, depending upon your purpose, time and equipment, plus the number of plants you are going to put out.

There are 3 basic systems or patterns to which plants are set and allowed to spread: 1). the hill system, 2). the matted-row system and 3). the spaced matted-row system.

In the hill system, plants are set close together and runners are pruned off or prevented from rooting new plants by the use of plastic film mulch. Plantings are usually made in double or triple rows, in which plants are placed 12 inches both ways from each other with a 20–24 inch alley between each double or triple set. In a double-row hill system, an acre accommodates 29,000 plants; in a triple-row hill system, 32,670. That's one of the secrets to those big per-acre commercial yields—high-density plantings.

The matted-row system is the opposite extreme of the hill system. For the matted row, plants are spaced in a single row, 18–24 inches apart, with about 42 inches between rows. An acre takes 6,225 plants at the 18 inch spacing, 8,300 with the 24 inch spacing. The plants are allowed to send runners out to make new plants willy-nilly, though the grower will try to keep a 14 inch alley between rows for the pickers to walk in.

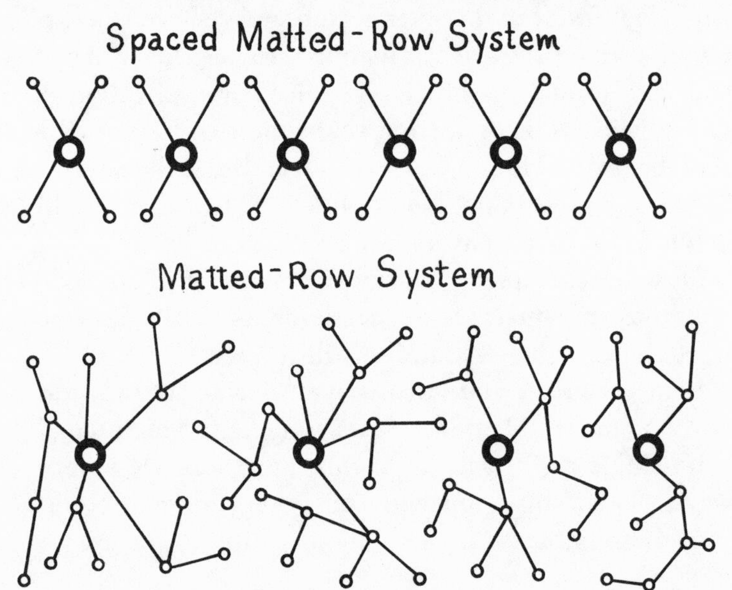

The matted-row system is the easiest one to maintain—it takes less labor. But yields will be less than in the double or triple hill system, and the quality will be lower than in the spaced matted-row system.

The spaced matted-row system, as the name implies, is a growing method intermediate between the other 2. Plants are allowed to set a *few* runner plants, but not many. Most gardeners with small patches of berries favor this system. Like myself, they will allow 4 to 6 runners to root equidistant around the mother plant and keep other runners pruned out. Keeping those runners pruned is not an easy job, but in the next year your reward is very high quality berries. Spacing is about the same as in the matted row system, though you can put the plants closer together if you want to. I like the 18 inch spacing between plants.

The Actual Planting

Having decided upon your system and row spacing, use a hoe to mark the holes where you will set the plants. One chop at each location is enough to mark the spot. (Of course, if you are planting a large plot with a mechanical transplanter, you don't need to mark with a hoe.)

Now get the plants from the refrigerator, unpack them and drop them into a bucket of water. If you have ordered more than 1 variety, each variety will be marked and tied separately. Untie only 1 variety at a time so you don't get them mixed up in the bucket. Allow the plants to soak in the water from 15 minutes to a half-hour, then let them stay in the water (keep the bucket in the shade) until you are ready to plant them. The water taken up by the plants aids them in withstanding the shock of transplanting. More important, the roots won't dry out while waiting to get into the ground. If you leave plants in the sun and their roots dry out, bad news.

With the plants from every good nursery come directions on how to set them in the soil—neither too deep nor too shallow. And if the directions are not included with the plants, nearly every garden book or plant catalog will show you how—with pictures. Indeed, the fact that the soil line on a transplanted strawberry plant should come up to the exact spot where roots and crown meet is the most publicized piece of garden how-to ever committed to paper. To my way of thinking, the planting depth is not really all that critical. Just use your head: Don't leave the plant so high in the ground that you can see the tops of the roots and don't bury the crown under the soil. Simple.

Just as important as depth is the way you should position the roots in the ground—but people who write books with one eye on other books rather than on experience don't mention that so often. Here's what I mean.

You've made a depression in the ground with the hoe previously to mark your spacings. You have the strawberry plant in your right hand. Put your left hand into the depression made by the hoe and with palm down form a ball of earth in the bottom of the depression about the size of a tennis ball. Nothing elaborate—just squeeze a fist-full of dirt together. Now place the strawberry plant on top of the ball of earth and push it down *gently* with a very slight twist. The little mound of dirt plus your twist will spread the plant's roots out into a fan shape. That's the way roots of any plant should look before you put the dirt on top of them—*spread out almost horizontally, not bunched together going straight down.*

Now cover the roots, still holding the plant in your right hand and brushing dirt in with your left. As you learn to coordinate both hands to this task, your right hand will almost automatically adjust the plant up or down a little bit so that the roots are covered adequately but the crown remains above ground. But always push the dirt around the plant a little higher than it should be, *then press the dirt down solid around the plant.* If you just brush dirt loosely around the plant to the proper height, any hard rain will wash the dirt away and expose the roots.

Now I hasten to add that you may very well get a good stand of plants without going through this elaborate method. If you are putting in a great number of plants—with machine or by hand—you won't have time to be that careful. You'll sock the plants in as quickly as you can, and accept the risk of losing some. But on a small patch, where losing a few plants is more critical, painstaking efforts will pay off. Not only will the plant get a better start, but when roots are spread out more or less horizontally in the ground, the plant makes a firmer anchor that is more difficult for frost to push up and out of the soil during winter-spring thaws.

I usually trim the roots a little before I set the plants in the ground. Some authorities do not recommend root pruning of strawberry plants, but I keep on doing it because I have always had almost 100% survival of my plants. It's a very simple operation. If you hold up a plant, or better, a handful of them, the roots hang down looking a little bit like the beards so many men are sporting these days. Take a scissors and trim off the scraggly ends—about an inch's worth will do it.

Incidentally, your plants at this stage may look dead to you, or at least quite forlorn. Some have a few leaves, some hardly any. The appearance means nothing as to the health of the plant. Any leaves present are last year's and will wither away as new leaves start to grow. Judge the plant by its root growth. Usually your package will contain a plant or two that has hardly any roots. That's why your package will invariably contain a couple extra plants. If a plant has very little root growth, plant it with a prayer but don't expect much—even if the good Lord is on your side.

I do not plant on *raised* beds as I've seen some folks do. I do not like raised beds—ground built up above the natural level of surrounding soil—because in dry years, plants on such beds usually suffer a greater lack of moisture. However, I shape the alleys between the rows about 5 inches deeper than the prevailing soil level—for 2 reasons. When you are picking the berries, having your feet (or knees) even 5 inches closer to the berries helps your back a little. Secondly, in case of very wet weather, the water will stand in the alleys, not in the rows inundating the berries or plants.

Plant Maintenance

All right. Your berry plants are in the ground, and your back needs a good rest. You stand there at the edge of the garden, and you can barely see the object of your solicitude. But within a remarkably short time (4 days for me if the weather is reasonably spring-like), new leaves will begin to form, and in a few more days you'll be able to "row" the plants by eye across the patch.

First comes a set of 3 new leaves emerging from the crown. A second 3 leaves appear as the first set opens to full size. Then a third. About that time, some 25 to 30 days after the first leaves appeared, the first flower stems branch out from the crown. The buds will flower and set berries if you allow them to. If you have a small patch, you may want to take the time to pinch off those blossoms as soon as they appear so that the vigor of the plant goes into making more leaves, roots and runners for next year's crop. The job is too laborious and time-consuming on a large planting, but whatever blossom pinching you do will be rewarded by a bigger and better crop the following year. To "pinch" off a blossom, you squeeze the blossom stem—or better, the whole flower stem of several blooms—between your thumbnail and middle fingernail, cutting the stem in 2. Never try to pull a blossom or flower stem off—you might pull the whole plant out of the ground.

More leaves will continue to emerge from the plant crown all summer. Also, shortly after blossom time, the first runners will begin to shoot from the crown, and more runners will grow out all through the summer.

When a runner's tip has grown about 8 inches or more away from the berry plant, it will bend upward, almost at a right angle and begin to form a new plant. At the bend of the angle, or elbow, roots will grow out and into the ground, and the new plant is started. Once rooted well and growing, the new plant will send out its own runners. In this way a strawberry patch spreads and multiplies.

If you are training your plants to a strict hill system, you will prune all runners. If you are committed to some variation of the spaced matted-row system, you want a few runners to root and make new plants, but the rest you will prune.

Generally speaking, a runner plant will produce only about one-third the berries that a mother plant produces. But you can get better production from runner plants if you get them rooted as early in the growing season as possible. This will take some active help from you, especially if your plants are mulched thickly. Runners take a longer time to put down roots through mulch, and if the mulch is over 5 inches thick, the runners might not root at all.

Therefore, when a runner tip has made its 90-degree bend upward, I push the elbow of the runner down through the mulch so it is in direct contact with the dirt. To hold it there, I may lay a small pebble on the runner if one is handy or dig a slight depression in the soil with my finger, place the runner elbow in it and compress dirt around it. The runner roots quickly in this position and still has plenty of summer growing time to develop a good crown for higher berry yield the next year.

I allow 4 runners to root around the mother plant. All the rest of the runners I try to keep pruned. On a patch of 100 plants, it takes about an hour's work every 2 weeks to keep runners under control. You'll make most of this time up the following year because pruning insures that a much greater portion of your berries will be large ones. Therefore you'll be able to pick a quart of berries very quickly.

However, if you don't want to bother with runner pruning, you can let your plants grow to the matted-row system and still get a fairly good yield, all other things being equal.

About 2 weeks after you've set out your plants, they should be growing well. You will want to cultivate once or twice to kill germinating weeds. Your only other worry (at least *my* only other worry) is rabbits. Rabbits will eat a few of those first new leaves on a new transplant. They won't touch old leaves in an old patch, and rarely eat even new leaves in an old patch. But they like new leaves of new transplants—plants which can least afford the loss of new leaves. Shows how cussed rabbits are. They'll do the same thing on new raspberry canes. It's a mystery to me. I do not like wild rabbits at all; I am Mr. MacGregor personified. My advice on controlling rabbits is to shoot or trap them and make rabbit stew.

This advice is repugnant to many people who think they are sturdy environmentalists. All I can say is that I can prove with mathematical precision that if someone or something doesn't decimate the rabbit population (and rabbits have no natural enemies left in outer suburbia) the poor dears will increase and multiply astronomically, finally run out of gardens to plunder, become malnourished, get diseased and die anyway. And perhaps pass a little rabbit fever—tularemia—around in the bargain. Since I've already said that the best place for a wild rabbit is in a stew, let me add, in qualification, that only healthy wild rabbits make good stew. Sick wild rabbits may infect you with rabbit fever.

At any rate, rabbits won't eat many of your plants because soon your peas will be up, and they'll eat pea plants instead. Sometimes—most times—the strawberry plant will recover if its first set of leaves is eaten off. But if a second set goes the same way—caput.

If rabbits have access to new red clover in spring, they will eat it, sometimes in preference to garden things. I found that out quite by accident and pass it on only hesitantly. Next year the rabbits may decide to avoid my clover like it was poison ivy.

Mulch

After the plants are growing well and before real warm weather arrives, I mulch the whole strawberry patch after giving it a final thorough cultivation. The mulch should snuggle up next to the plants without covering them. Strawberries like cool moist growing conditions, which the mulch helps provide during hot weather. Not everyone mulches their plants for the first year of growth. I do, because mulch is equal to at least one good rain or soaking irrigation during dry weather. And mulch is cheaper than irrigation.

For the first mulch, I use strawy horse manure 4–6 inches deep. Since no berries will be picked off the plants this first year, the amount of manure in the mulch is of no sanitary (or unsanitary) consequence. (Mulch on my bearing, second-year plants will be clean straw or pine needles, but I don't want to get ahead of myself.) It is no problem for me to get horse manure as there are many horse lovers in the area whose loving horses produce more manure than their loving owners know what to do with. Any strawy animal manure, except rabbit or chicken manure, which I think is too rich in nitrogen for strawberries, would work just as well. But I prefer horse manure, not for any especially good reason other than that it is available and has always given me good results. I *do* have a weakness for folklore, and there *is* an old folk saying that goes: "Wild strawberries follow horses." Well, that may mean absolutely nothing, but *if* wild strawberries once grew well where the farmer's work horses left their droppings, then why not tame strawberries too? A little folk whimsy never hurt any man's garden.

If you can't get strawy manure, use fresh grass clippings 6–8 inches thick; fresh clippings have some fertilizing value to them in addition to their mulching value. Fresh grass clippings from fertile soil contain as much as 6% natural nitrogen which becomes readily available to the plants as it rots.

Once your mulch is in place, most of your weeding chores are finished for the year. A few weeds will grow up around the plants where you can't spread the mulch thickly for fear of covering the plant. Your strawy horse manure will probably have a few grains of wheat or oats in it which will germinate and grow. These will have to be pulled along with a few stubborn weeds that always manage to push up through mulch no matter how thickly you spread it.

Through the summer, your only steady task is keeping runners pruned or otherwise under control. By December, most of the summer mulch has rotted away—at least in this humid area. As soon as the ground freezes I cover the whole berry patch with another 6 inch layer of clean straw (baled from the field; not straw used as bedding in a stable) or pine needles or a mixture of leaves and pine needles. I've used the latter exclusively the last two years—I get it free from neighbors while straw is becoming increasingly expensive to buy.

Covering strawberries after the onset of cold weather is important. The purpose is not primarily to protect the plants directly, but to keep the ground from alternately freezing and thawing, the way it would if exposed to the weather. When the ground freezes and thaws continuously through the winter as it often does in our part of the country, it will heave strawberry plants out of the ground. But under a heavy mulch, the ground, once frozen, will usually stay that way most of the winter. When it does thaw in late winter or early

Elimination of weeds is a major reason for mulching, but the thick layer of an organic mulch, such as the hay Carol Logsdon tucks around each strawberry plant, also conserves moisture, adds to the tilth and fertility of the soil and keeps the berries clean.

spring, usually the weather won't get cold enough to refreeze the ground under mulch.

With the first warming of spring, your strawberries, asleep under their blanket of mulch, begin to stir. Don't be too hasty to uncover them. As I have said, I keep the mulch on as long as possible, because I don't want the berries to start growing any sooner than necessary. Earlier berries are the ones most in danger of blossom-killing frost. But don't wait too long either. Once new growth starts under the mulch, get the mulch off, or it may smother the plants.

Pull the mulch between the rows of berries where it will make a nice clean path for pickers, keep the berries free from dirt and control weeds. Organic berries ripening on top of mulch (rather than dirt) never need to be washed before eating.

Until the berries form, you have nothing else to do except worry about frost. If a frosty night is forecast when the berries are blossoming, you can cover small patches with old blankets, sheets of plastic film, anything like that. Be sure to pin down the covering with rocks or other heavy objects or the wind might blow it off. Once the blossom petals fall and the new berry begins to form, danger of harm from frost has passed.

Commercial growers turn on sprinkler irrigators to fight frost. The water raises air temperature around the spray area and keeps the frost off the blossoms. You could do the same thing on a small area with a hose set to a fine spray. Wind is a good frost fighter too, sometimes keeping the air in motion enough to hold the temperature above freezing. Some orchardists install wind machines above their trees. Others have hired helicopters on critical nights to hover above an orchard and keep air moving. That is a very expensive way to fight frost. I have not heard of anyone setting big window fans in small gardens on frosty nights, but it certainly makes as much sense as using helicopters on orchards.

As your berries begin to ripen, it won't hurt to put a scarecrow in the patch and start feeding your cat there too. A scarecrow may seem a bit ridiculous, but if it keeps away only 5 birds, you've gained another box of berries. If moved from spot to spot every other day, and if it has some movement of its own in the wind, a scarecrow will scare away some birds some of the time. Pheasants are easiest to scare away in my experience. They seem more fearful than crows, robins and blackbirds. If you have brown thrashers, they will surely eat a few berries, but that's well worth the joy of having such astute companions around.

The Harvest

Once harvest season begins, you'll want to pick your berries every day, or at least every other day. I prefer picking in the morning after the dew is gone, or in the cool of the evening. In a separate basket, save old, bruised and partially rotted berries for jams and jellies. Pick all ripe berries as you go—don't skip around picking just the large ones.

You may have trouble getting your work finished during strawberry season because suddenly friends you hardly knew you had will just happen to stop by for a visit.

Many people do not know how to pick strawberries. They try to get the berry off without the stem, and in so doing usually squeeze too hard on the berry, bruising it. Pick a berry with the stem on it. To pick a strawberry, grasp the stem right above the berry between thumb and forefinger, and pull gently while at the same time cutting the stem between thumbnail and fingernail. Some varieties almost pop loose by this method—Marlate is excellent for easy picking. Other varieties "pick hard"—one of, if not the only, weak points of the fairly new variety Guardian. If you try to pull a berry loose without the cutting action of your fingernails, you will likely pull a whole cluster of berries loose, most of them not yet ripe.

I know I don't have to tell you how to eat strawberries. Try them with a light table cream if you haven't—and clover honey. So what if you do gain a little weight—you can work it off easily enough in the strawberry patch.

Maintaining the Garden

After harvest, you have another decision to make—whether to save the patch for another year or to plow it up. If you did not start a new patch this spring, you will most likely want to keep your present one for another year. Many growers mow the strawberry plants down immediately after harvest as the first step toward maintaining the patch another year. Others will thin the patch by going through and pulling out all the old mother plants. Still others will tear the living bejeebers out of the patch with a row cultivator, getting the matted patch back into some semblance of rows. I recommend mowing the patch, then either by hand or with a runner cutter, trim the boundaries of the beds back to almost the original row.

That is not an easy job, and it is one more reason why I do not try to keep a patch for a second bearing year. Even with careful renovation, the patch usually will bear only about half the first year's yield in the second year. The patch will become weedy. Disease is more liable to creep in. And while the patch is not doing well in that second bearing year, it is taking up valuable space in the garden that could be more profitably used for something else.

So after berry harvest, which ends about July 2 here, I shred the plants with my rotary mower, work the shreddings and roots into the soil with my tiller, and immediately plant sweet corn or soybeans or fall vegetables. That land is thus fully utilized by doublecropping. From then on my strawberry work is confined to the planting I made this spring, which will bear well next year.

It is possible to set strawberry plants in late summer or early autumn for bearing the following spring. That shortens the time from planting to eating considerably, but you won't get as good a yield—at least not in the North. And you will have to have plentiful rain or supplemental irrigation to pull it off. Plants kept dormant in cold storage will usually respond better in late summer plantings than plants moved from an old patch. Pocahontas and Albritton are supposed to be good varieties to try for late summer planting. Farther north, in climates colder than these varieties usually grow in, late summer setting is not advisable. Cold weather can kill late set plants. Even if your winters are not severe, play safe by covering at least your late set plants with mulch.

Though I believe in buying new plants each year from a nursery, you don't have to do that. In the spring, you can transplant year-old runner plants from an established planting and grow a fine crop. I hesitate to follow this practice very much because I'm afraid of bringing in some disease that had just started to infect the old plants. But if you have found you have some particular plants which yield an especially nice berry, you should by all means save some runner plants from them. I'm convinced that I have had strains of certain varieties which were of higher quality than the ones I'm buying now. I wish I'd saved those strains. I'm pretty well convinced —and so are some other gardeners I talk to—that strawberry variety strains do run out, even in nursery plantings. But of course, that would hold true eventually for your own plants even more than for nursery plants.

Diseases and Blights

I haven't said much about the diseases, blights and bugs that can harm strawberries. I always hesitate talking about these subjects because any such discussion can scare a beginning grower to death. So let me preface the bugs and diseases by saying that in all the garden patches of strawberries I have raised, and that's a lot, only one disease, gray rot, has been a problem for me—and then only once a serious problem.

To control strawberry diseases (or any berry diseases) the grower should follow a good sanitary program. Step No. 1: Buy disease-free stock from reputable nurseries that sell plants monitored by state plant inspection services. Let me qualify immediately. You can buy total security nowhere, no how, from no one. If you read the certificate of inspection given to state-approved nurseries, you will find it says something like: ". . . said premises and nurseries were found to be *apparently* free from dangerously injurious insect pests and plant diseases." My italics on the "apparently." When you buy plants from a reputable nursery you're not buying absolute protection. But if you buy from an uncertified source, you are taking a bigger chance. And if you transplant runners from an old run-out patch in a neighbor's garden, you're taking an even bigger chance.

The "virus-free" program for strawberries has been in effect for quite some time now and has been so successful that virus disease in strawberries is no longer a serious problem. Almost all strawberry plants sold in this country are now "essentially" virus free and can be certified "Registered" VF plants. That means the plants have been grown from stock provided by the USDA (generally working with one of the land grant colleges). The original stock is grown under very rigid sanitary conditions.

But since viruses do not usually produce clear-cut symptoms, and, furthermore, since plants can be re-infected, there is no absolute guarantee that your plants are 100% virus free.

But again, it's safer to buy registered plants than unregistered.

Sanitation Step No. 2: Buy varieties that grow vigorously in your area. A healthy, vigorous plant is less apt to be badly harmed by any enemy.

Sanitation Step No. 3: Rotate your berry patch regularly. I plant strawberries on the same ground no more than once out of every 4 years. I keep a patch only 1 bearing year rather than let it linger to encourage bug or disease buildup.

Sanitation Step No. 4. Buy varieties resistant to diseases prevalent in your area whenever possible.

Red stele is supposed to be the most serious fungal disease of strawberries in the United States. It is also supposed to be common throughout the northern two-thirds of the country. I guess I'm just plain lucky never to have been faced with it.

The name derives from the surest symptom of the disease: a reddish-brown root core rather than a normal yellowish-white core. The part of the root surrounding the core retains the healthy color, so it's easy to spot the abnormal red center. The red may extend the length of the root or it may show only a short distance above the end tip.

A reddish hue in the crown of the plant is not a sign of red stele—nor are red leaves, which can be caused by a number of nutritional irregularities. A few red leaves in late summer and fall is no more cause for worry than red leaves on an autumn maple tree.

Red stele stunts plants, which wilt and sometimes die just before fruit starts to ripen. Symptoms sometimes disappear in warm, sunny weather. The fungus, as is true of most fungi, is most active in wet weather. There's no cure for red stele. It will not persist in well-drained soils but it can spread from poorly drained land to other land, carried by transplanted diseased plants, by soil on machinery or by water moving off of infected soil.

While there's no cure for red stele, some varieties are resistant to it. In the East, that's Guardian, Redchief, Sunrise, Surecrop, Sparkle and Midway; in the West, Hood, Molalla, Shuksan, Totem and Siletz. Experts say that new races of red stele have appeared which can damage some of the resistant varieties, but those mentioned are your safest choices.

Look your roots over carefully when you are putting plants in the ground. A root that looks like a rat's tail—any root without rootlet branching—may be an indication of the disease. State inspectors may not have spotted the disease because it may not have been present when they looked at the plants in the field or because only a few plants were infected. Cut into suspect roots to see if the distinctive reddish center is present.

Verticillium wilt is another disease that can attack strawberries, among other crops. The fungus occurs throughout the U.S. and is most active in cool, humid weather. I think this wilt killed a couple of my plants this year. One of its odd characteristics is that it may appear in a mother plant but not in rooted runners from that plant—which is what happened in my case. But here's the Department of Agriculture's description of verticillium on strawberries—in case it helps you more than it does me: "Outer leaves wilt and dry and the margins become dark brown. Few, if any, new leaves develop. New roots that grow from the crown often are very short and have blackened tips. Plants appear to be dry and flattened. Black sunburn lesions may appear on leafstalks and runners. Severely affected plants collapse, sometimes abruptly. Less severely affected plants usually recover and produce normally the next year. In the West, affected plants usually do not recover."

I can make at least one deduction: we Easterners are luckier this time than the Westerners.

Special sanitary precautions are in order if verticillium wilt is prevalent in your area. First, don't plant strawberries on land that was growing tomatoes, peppers, potatoes or other strawberries during at least the last 2 years. I violated that law this year, putting my strawberry patch where I had tomatoes last year. This could explain why I lost two plants.

You can never tell just by eye the cause of a wilt. Not even experts can. But it is fairly easy and gives a marvelous impression to stand in front of someone's wilted plant, rub your chin sagely and say, "Verticillium wilt, probably."

In the West, don't plant strawberries after the above mentioned crops or where cotton, okra, melons, eggplant, mint, apricots, almonds, pecans, cherries, avocados, roses or cane fruit grew anytime in the 10 previous years, USDA says.

Varieties that are somewhat resistant to verticillium wilt are: Blakemore, Catskill, Guardian, Marshall, Redchief, Salinas, Robinson, Siletz, Sunrise, Surecrop and Vermilion. Varieties most suceptible to verticillium are: Dabreak, Dixieland, Earlidawn, Jerseybelle, Klondike, Lassen, Molalla, Northwest, Raritan, Shasta and Vesper.

Worms and Rots

Nematode infection in strawberries cannot be easily cured either, but it is 1 case where nearly everyone agrees that good organic practices are the best preventive. Nematodes are very small worms that live in the soil. Unless you know what you are looking for, you won't see them with the naked eye. Several kinds can harm strawberries: the root-knot nematode, the root-lesion nematode and the sting nematode (the latter only in the Southeast, especially Florida).

The root-knot attacks strawberries in the northern two-thirds of the country; the root-lesion is distributed all over. The root-knot forms swellings on roots called galls which are about an eighth-of-an-inch in diameter. Just above the gall there will usually be several short branch roots. The root-lesion nematode causes roots to turn wiry and brown. In both cases, plants are weakened by severe infection. They become less vigorous, especially if the weather turns dry, and yields are much lower than normal.

Nematodes seem to prosper in sandy soil but not in clay. And they rarely amount to much in fertile soil, high in organic matter. Maintaining a high level of organic matter in the soil, the experts agree, may prevent or at least reduce nematode damage. Chemistry's answer is fumigation, costing $500 an acre or more.

Frequent but very shallow cultivation helps keep infected roots from further damage.

Marigolds are another nematode fighter, as organicists know—and that isn't just fanciful folklore either. Scientists have found that marigolds do indeed produce something that nematodes don't like. If your soil is infested, plant your future berry patch to marigolds for a year. This method is cheaper than fumigation and a sight prettier.

Half-a-dozen fruit rots do damage to strawberries, and the worst, if you ask me, is gray rot. I've already given some organic control methods, and they apply to all the fungal rot diseases: Apply fertilizer the first (non-bearing) year and go easy or apply none at all the spring of the fruiting year. Mulch. At least don't cultivate from blossoming to harvest. Pray for dry, sunny weather.

Just for the record, I've tried some of the non-organic fungal sprays—Captan, Ferbam, Zineb—and I agree with what the government bulletin says: "This control measure (fungicides) on strawberries often costs more than it's worth."

Insects

Some 24 kinds of bugs attack strawberries, but don't be overly alarmed. Not a one of them has seriously affected my strawberries in over 10 years of not spraying insecticides on them.

First there's aphids—there must be aphids especially designed to harm every plant! On strawberries, they are the leaf aphid and the strawberry-root aphid. The chief harm of leaf aphids, however, is their capacity to transfer virus diseases from old infected plants. If you have none of the latter around, there's not much of a threat. The strawberry-root aphid sucks juices from the plant via the roots. However, the blame falls on the cornfield ant, which herds aphids about like a dairyman herds cows. The ant carries the aphids below ground to the roots. The best way to control the aphid is to control the cornfield ant by frequent cultivation.

Two kinds of mites can harm strawberries, the cyclamen mite and the spider mite. Organicists will find remarks from a USDA bulletin on mite control particularly interesting: "Cyclamen mites may be killed by hot dry summers in some areas. Sometimes they are kept under control by predatory insects. The use of parathion or malathion to control any pest may also kill the beneficial insects that destroy mites. Then there will be a rapid buildup of the cyclamen mite population. Parathion or malathion should not be used on strawberries where the cyclamen mite occurs."

Reminds me of what an entymologist told me several years ago: "As far as mites are concerned, I sometimes think we'd be better off if we'd never tried to spray them in the first place . . . but don't quote me."

There are two families of root weevils whose larvae feed on roots and crowns of strawberry plants and sometimes cause serious damage. Organically, your best control is to plow up the patch immediately after the first harvest, as I do, to prevent weevil buildup. The same control works for the strawberry crown borer. Since the larvae do not change into adults until fall, shredding the plants and plowing up the patch after harvest will kill many borers. Also, since the adult of this species cannot fly or crawl more than 300 feet, say the experts, locating a new patch more than 300 feet from old ones and from wild berries will help prevent infestation.

The strawberry weevil leaves a trade mark you can easily use for identification purposes: adults sever stems of fruit buds but not completely. You'll see the stems hanging by a thread.

Other insects that cause only minor damage, but which may become serious when they experience a sudden population buildup, are white grubs and wireworms. If you don't plant strawberries on land that was previously in sod, you won't ordinarily have trouble from grubs. Another minor problem in the Pacific Coast area is the awesomely named omnivorous leaf tier. The tiny worms of the larval stage live in crevices of rough-barked trees over winter, and the best way to avoid them is to not plant strawberries near rough-barked trees.

Flower thrips, lygus bugs, mole crickets, potato leafhoppers, field crickets, flea beetles, snails, slugs, spittlebugs, leaf rollers, strawberry rootworms and strawberry whiteflies will occasionally give you fits, especially the last. The minute moths are about one-sixteenth of an inch long and covered with a grayish-white powder. The larvae ooze a sticky substance onto the leaves as they suck out plant juices.

I'm not making much of an attempt to describe bugs because I know from experience you can't possibly identify a bug on a plant with a verbal description alone. For instance, here's the entymological description of the strawberry rootworm: "Adults are oval beetles; shiny dark brown or black; some have four dark spots on wings; about one-eighth inch long. Larvae are small, white, brown-spotted grubs; they feed on roots of strawberries in early summer." A good, helpful description. But you need a picture with it.

I urge every gardener, especially every organic gardener, to become a student of insects—an expert bugwatcher. You need to know as much about the habits and cycles of the bugs in your garden as you know about your plants. Get a couple of illustrated bug books for your library. Not only will you become a better gardener, but you just may begin a fascinating hobby of collecting bugs or photographing them or just watching them. Bugwatching is easier than birdwatching and more exciting than a bullfight or a ballgame. Ever watch a praying mantis prey? A large moth emerge from a cocoon? An ant "milk" an aphid? A wasp capture a spider? Have you tried to duplicate the leafhopper's fluorescent greens and purples on canvas? Caught a firefly glowing on film?

Understand your garden's insects, and you can insure your garden's bounty.

Strawberriana

Strawberries can be adapted to what I call "novelty horticulture." You've seen at garden stores special "strawberry pots" with cupped openings around the sides? You start the mother plant in the pot and as the runners grow, set them in the side cups. Strawberry barrels work similarly. Fill a pot or barrel half-full of gravel, then complete filling it with good top soil. Fish emulsion is a good organic fertilizer to use for potted strawberries.

I have always wondered why (or if) growing strawberries in pots, barrels or tiers was so popular, since it is hardly an efficient way to put strawberries on the table. But if your only garden is a patio or an apartment balcony, it's 1 way to get a few handfuls of fruit. Certainly this is one situation where an everbearing variety would be a good choice to try.

I've seen "climbing" strawberries advertised, and people ask me if they are any good. I have visited only one gardener who tried them. She found the berry an interesting novelty, but yield was only fair. "Before I learned to take them down off the trellis and cover them with mulch, they froze and died in the winter," she told me.

Here's a few more ideas about strawberries, so you'll know

you "read them somewhere, but can't remember where." Strawberries make a good wine. Strawberry-rhubarb pie is not hard to take either. A facial cream using strawberries and oatmeal was once thought excellent for the complexion. Juice strained from mashed ripe strawberries was mixed with vinegar and used for the same purpose. It seems a waste of good strawberries to me.

If you make homemade ice cream, you probably already have a good recipe for strawberry ice cream. If you don't—or even if you do—try this way once. Make vanilla ice cream (the kind for which you do *not* cook a custard beforehand). When the ice cream is about three-fourths made, open the freezer and throw in some ripe berries, sliced well (about a pint per 3 quarts of ice cream). Then finish making the ice cream. That's livin'!

Some of the cake-textured shortcakes served with strawberries, especially in the North, are okay I suppose, but I have a hunch their main purpose is to make the strawberries go a little farther around the table than they would if the family didn't have the cake to fill up on. If you grow your own strawberries, you don't have that worry. You have plenty of berries. So try a shortcake that does not compete with the fruit.

The best shortcakes are made with biscuit dough. (No, you don't have to agree with me, but try it first before you argue.) Sift together a cup of flour and a teaspoon-and-a-half of baking powder, one-fourth teaspoon of salt and 2 tablespoons of raw sugar. Cut in 4 tablespoons of butter or shortening (we use butter). Add one-third cup of milk. Roll out into a big biscuit and bake at 450° until browned. Serve warm from the oven.

Cut yourself a piece about 5 inches square and cover it with fresh berries, sun-ripened on the plant. Do not sprinkle the berries with sugar, or sugar syrup. A good ripe Fairfax needs no sugar. Just a dribble of clover honey over your serving and

By picking time, the layer of mulch, now slowly decomposing into the soil, is overshadowed by lush, healthy strawberry plants, heavy with berries. The mulch has kept the spaces between the rows tidy, and the pickers—here the author and his son—have easy access to the berries.

then a generous amount of light cream. Now you have something before you no gourmet, however wealthy, can improve upon.

Chapter Three

The Priceless Raspberry

I like to refer to the raspberry as the proverbial "pearl without price." It is easier, in most sections of this country, to buy a Rolls-Royce than a fresh raspberry. Raspberries just ain't for sale regularly except in places like Washington and Oregon where the fruit is still grown extensively on a commercial basis, or in classy gourmet shops where you may pay as much as *$1.25 a half pint* for them.

Raspberries have fallen victim to a disease of modern progress: they "don't pay." "Don't pay" means "a lot of trouble when there are easier ways to make a buck." Grandfather's old harangue has become only too true—nobody wants to work anymore. Ever since the assembly line became the handmaiden of economics and successfully divorced the working man from responsibility for the product of his work, we have only *seemed* to gain a better quality of life. Our assembly line bread lacks nutrition, our assembly line clothes fall apart, our assembly line houses buckle and leak, and our assembly line automobiles keep us too poor to pay for houses and food and clothing that do have some quality to them.

Like homemade bread—or anything else worth enough to

The raspberry offers several nutrients: vitamins A, B_1 and B_2 and calcium, phosphorus and iron. The real reason they are so popular, when they can be had, is their delicious flavor.

demand a little extra work—the raspberry won't bow to the assembly line process. At least not yet. It has to be ripe or almost ripe before you can pick it—so growers can't harvest it green and ship it to far away markets ahead of the season. Once ripe, the raspberry is quite perishable and must be sold within a day or 2 after picking. Planting, pruning and harvesting raspberries remain to a great extent hand chores—machines can't do any of these jobs very well yet. Raspberries "don't pay."

All of which should please rather than displease you, the gardener or small farmer. Knowing how rare good, fresh raspberries are will make yours taste all that much better. Also the "don't pay" situation spells opportunity for you. You can be assured that you can sell your own berries locally without much competition from commercial outlets. Mr.

American Bigness hasn't found a way yet to muscle in on the fresh raspberry market.

The raspberry is also a good practical crop for organic gardeners and farmers because it will produce satisfactorily with organic methods. Good production depends mainly on adequate moisture, proper pruning and careful sanitation to avoid virus and other diseases. And the labor involved in a raspberry planting is not all that tremendous—far less than what you expend on the same acreage of strawberries.

Raspberry Family Traits

There are two rather distinct kinds of raspberries—2 families, to say it more correctly. The red raspberry and its sister, the yellow raspberry, make up one; the other includes the black and the purple varieties. The reds and yellows have a different habit of growth than the blacks and purples. I shall try to treat them separately—first the red family, and then the black family.

The stalk of a raspberry is referred to as a cane. Red and yellow raspberries multiply and spread by sending up new canes from the roots of old canes. If you walk into a patch of red or yellow raspberries during June, you will find canes that are bearing fruit or about to bear fruit. These canes will be hard and turning brown, while the canes that do not have berries are green and succulent. These green, softer canes are the new ones that will have berries *next* year (or in the fall of this year, if the variety is a fall-bearing one). The canes that have berries forming on them now were green and succulent last year, and as soon as the berries ripen, these canes will begin to die.

Once you understand this growth habit, you know how to handle the plant. Just remember that every year new canes come up from the roots of the old canes, and that every year the canes that came up last year will bear fruit and die. (With fall-bearing varieties, the cycle is not quite like that, but I'll discuss these varieties by themselves a little later on.) In a healthy, vigorous patch, more new canes come up each spring than you'll want to grow. When the new canes come up, they are usually referred to as "suckers." If the suckers are too thick in the row, or if they emerge beyond the boundaries in which you want to keep the row contained, you have to pull them out, as you would a weed. Or, if you want to start a new row someplace else, you can dig up the healthy suckers carefully and transplant them.

Planning Your Garden

With those few generalities firmly fixed in your mind, let's proceed to a detailed step-by-step plan to establish a producing raspberry patch.

First, select a variety to fit your area and climate. Neighbors who grow raspberries and nurseries who service your area can be of more help in this regard than I can. But here are some examples.

Latham is still probably the most widely adapted red raspberry on the market. It does pretty well in New England, New York, the Great Lakes Regions, the Upper Mississippi Valley and the Middle Atlantic States. Newburgh and Taylor will do well over most of this territory, as will Sunrise and Milton. Early Red is recommended for the Great Lakes too. On the Plains and Upper Mississippi Valley, Chief is the preferred plant because of its hardiness. In the Pacific Northwest, where raspberries are grown commercially, try Canby, Puyallup, Sumner, Washington and Willamette. In the South, where raspberries have not done well in the past, a new variety, Southland, shows much promise, according to plant breeders at North Carolina State University, who helped develop it. (Southland produces a fall crop too, and properly, I should not mention it until I discuss the fall-bearing varieties but, because it is the first variety developed that will bear fairly well in the southern Piedmont area, I want to get it mentioned here before southern readers skip to the next chapter.) New varieties come on the market almost every year—like Citadel and Reveille, lately from Maryland plant breeders. These 2 varieties are recommended for the Middle Atlantic States or anywhere where winter temperatures fluctuate above and below freezing a lot. A good variety in the Northeast for a late midseason crop is Ott's Pennridge. Hilton is another late midseason berry, considered to be the largest of all red raspberries. The oldest variety still around is Cuthbert, which was the universally grown raspberry before Latham came on the scene.

Of this group I've grown only Latham and Taylor. I like Taylor very much. Catalogs describe it as not needing support because the canes are so sturdy. Well, that's *almost* true. Some of the canes, if pruned back severely, will grow thick enough to stand erect without staking or trellising, but by the time the berries ripen, they often droop over to the ground. That is also true of several other varieties that are supposed to be "so sturdy they do not need support." If that turns out to be true for you, you are lucky. But whether staked or

A raspberry bush in bloom exhibits fragile, white blossoms.

drooping, Taylor is a good berry, conical rather than round, and good tasting.

Latham is okay too, if you get "virus free" stock. Virus disease in raspberries has reached a critical point, at least in the East, and virus-free programs are only in their infancy compared to the progress made in strawberries. But serious raspberry growers are turning more and more to paying out the extra money that virus-free plants cost. Those who don't feel that the money could be wasted because plants can become reinfected after a patch is established. But I strongly urge you when possible to buy virus-free stock, especially if you are purchasing only a few plants. As long as you're going to the trouble of growing raspberries, you might as well be as safe as possible.

Some nurseries offer 2 kinds of virus-free stock: Foundation Stock and Registered Stock. Foundation Stock is planted directly out of special screenhouses where the original virus-free stock was grown under complete protection from virus-carrying aphids. The Foundation Stock is sprayed periodically to keep aphids away and is grown at least 1,200 feet from all other raspberries. Plants taken *from* Foundation Stock plantings and grown elsewhere (isolated at least 1,000 feet from other raspberries) can then be sold as Registered Stock. The only place I know that offers Foundation Stock for sale is the New York State Fruit Testing Cooperative Association, Geneva, New York 14456.

Many commercial nurseries do not generally use the terms described above. Instead, they sell what they call "certified" plants and "substantially virus-free" plants. Certified plants are state inspected and certified free of diseases and insects—but not necessarily viruses, which often reveal no visible symptoms. Substantially virus-free plants are essentially the same as the registered stock described above. Virus-free plants are more expensive than "certified" plants. For five of the former the price in 1973 ran around $3.50; the latter, $2.00. The price differential shows up sharply when you buy in quantity: 500 virus-free plants sell for about $175. That amount of money will buy almost twice that many "certified" plants.

Working the Garden

Your plants ordered, you should now turn your thoughts to the soil you intend to plant them in. Actually, you would do well to begin preparing that soil at least a year before you purchase plants. It is very important that land into which you set raspberries be cultivated thoroughly beforehand. Oldtimers told me that, and in my wayward muleheadedness I ignored the advice. I tried to set plants in holes I dug in an old lawn. Half the plants died, and the other half never did do quite like they should have.

Raspberries like a humus-rich soil. Plant your future patch to buckwheat if the soil is heavy, to rye or clover otherwise, and plow the crop under for green manure. If your cultivating tool is a rotary tiller, mow the green manure crop first, letting the clippings fall on the soil and working them in with the tiller.

A plot of soil that has been in vegetables (except tomatoes or potatoes) for a couple of years and is in a high state of cultivation (by which I mean it is in good fertility and has been plowed, disked or tilled regularly) is a good spot to turn into a raspberry patch.

At any rate, work up the ground well before setting your plants. If your winters are not too severe you can set plants out in fall, but I advise spring planting whenever possible. When your plants arrive, you'll find that each one has about a 6-inch length of cane plus roots. That bit of cane makes a nice handle for you to grab while setting the plant. It will also help mark the row. But the plant will not grow from that cane; you want a new cane to come up from the roots.

Spread the roots out fan-shaped over a ball of earth just as I described for strawberries, only make the ball a little larger —about the size of a softball. Pull the dirt in level to the base of the cane and tamp around with your foot. As soon as new growth emerges from the roots, cut the old cane "handle" off —even if some leaves have started to develop on it.

You can mulch immediately, or cultivate around the new plants a couple of times first. At any rate, mulch before hot dry weather sets in for sure. Large growers can't always do that because it would take an awful heap of mulch to do the job properly on several acres. But I'm convinced that mulched raspberries—mulched every summer for 3 years or more—produce better than unmulched berries cultivated regularly.

Keep a 6-foot width between mulched raspberry rows. On a large planting, you may want to run a tractor and disk between rows for good, fast cultivation. If so, be sure you plan your rows wide enough to allow for machinery passage. The row of raspberries will grow—or should be kept—at not more than a 2-foot width.

In the row, set your plants about 3 feet apart. They will quickly fill the intervening space to make a continuous row of canes. Then it will be your job as pruner to allow only about 4 good healthy canes to grow per square foot of row.

Pruning and Trellising

The first year the only pruning you will need to do is to pinch off blossoms. You don't *have* to pinch them off, but the plants will grow more vigorously that first year if they don't have to put energy into producing fruit.

During the first winter, cut back canes to about 3 feet. The second summer, as soon as *new* canes start growing well, cut off the fruiting canes completely (the ones you pruned back during the winter). You don't have to do this; I advise it

An untended raspberry plant tends to grow into an impenetrable briar patch. Judicious pruning can ease the picking, and help control disease by opening the plant to better air circulation.

because I think the plant's energy ought to go into those new canes rather than into producing fruit so early in the plant's life. You won't get much of a crop anyway.

By the next summer though, the fruiting canes (which you pruned back to about 4 feet during the winter) will produce a crop—about half of what they will produce in another year or 2 when the patch has matured.

After the berries have been picked, cut the old canes out. Use a pruning shears and cut the canes off at ground level. At the same time cut out weak new canes, leaving only about 4 canes per square foot of row as I have said. In a vigorous stand, thinning may mean cutting out some perfectly good canes. Save the thickest. Allow no canes to remain closer together than the width of your hand.

Thinning allows better air movement through the row, but more importantly, bigger, high quality berries will result. Cutting out the old canes is a sanitary precaution, extremely important for an organic garden. The old canes may harbor bugs and diseases. Burn the old canes after cutting them out. You will be giving your raspberries as much protection as chemicals can give.

At the time you do your summer pruning (if not earlier) pull out all suckers that come up between the rows or too far out on the edge of the rows. Try to keep the row no more than 2 feet wide—foot wide rows are easier to weed.

Again come winter or early spring, you will want to trim back the canes to about 4 feet tall. I do, at least, because I do not trellis my red raspberries. Keeping the main fruiting

Red raspberries should be cut back to about knee height and thinned as in Figure 1 to maintain garden decorum (as well as ease harvesting, improve aeration and channel new growth). The old canes should be removed as production wanes. Figures 2 and 3 depict a black raspberry bush before and after pruning. The bush will probably need some support after the pruning is completed.

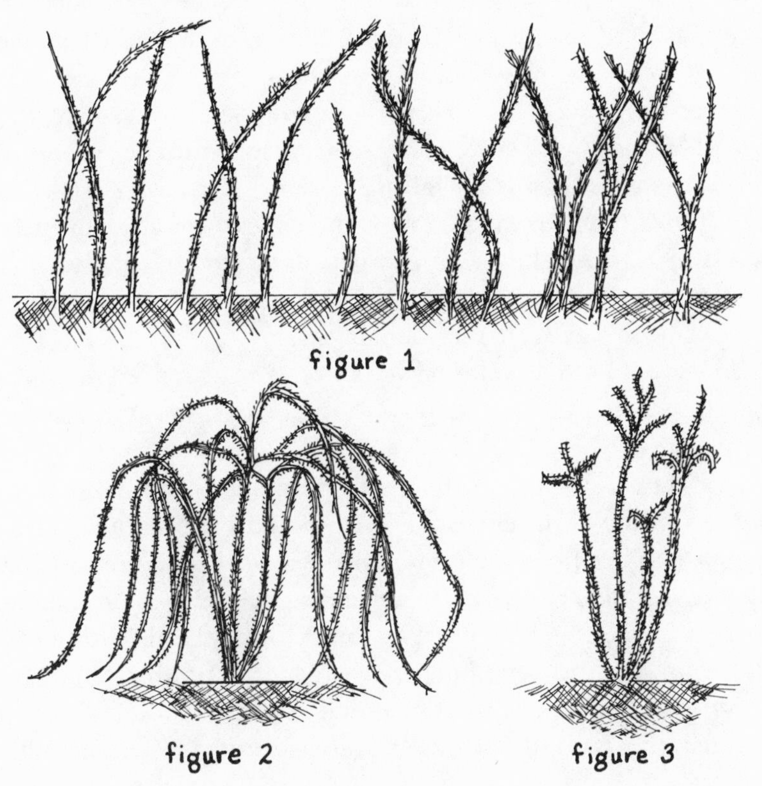

canes cut back helps them to support themselves a little better instead of falling over on the ground. If you intend to use some form of supplementary support, you can leave your canes at whatever height best fits your support, pruning back only growth that has winterkilled.

For example, one kind of training system for red raspberries keeps the plants in hills with 3 or 4 feet of space between hills. This system allows cultivation in both directions and reduces hand-pulling of weeds to an absolute minimum. Four canes per hill are usually allowed to grow. They are tied together loosely and trimmed back at winter pruning time to 5 or 6 feet tall. Tied together in a bunch, the canes are self-supporting.

Another trellis system consists of a 2- or 3-wire trellis much like that used for grapes. In the 2-wire system, the wires are strung over the row, one above the other. The canes are allowed to grow in a very narrow row, and are tied to the wires. In the 3-wire system, canes are slipped up between the wires to keep them from falling down. Either system makes a very neat and managable row and saves a great amount of space in a given plot—you can get more rows on it. But the system requires more hand work and more cost in posts and wire, and serves well for a small patch.

There are 2 other trellis designs you can use for red raspberries, but they are more adapted to the blacks, so I will discuss them later.

I neither trellis nor cultivate my summer-bearing red raspberries. Some of the canes fall over so that picking the berries from them becomes the same kind of stoop labor involved in picking strawberries. If I were selling the berries on a commercial basis, I would resort to some kind of trellis simply to speed up picking—the bottleneck in berry production.

As it is, I do not care too much that the canes fall over because the ground is thickly mulched and the berries will not get dirt on them. We never wash raspberries before put-

Trellising raspberry canes can be worthwhile. Here the canes are supported by a system of wires.

ting them on the table or before freezing—in my opinion, washing absolutely deteriorates the fragile fruit too much.

Maintenance

I keep my berry plants mulched permanently, adding new mulch each fall in the form of leaves, so cultivation is not a problem for me. If there's a lot of oak leaves in the mix, I sprinkle lime over them—just a little lime so the oak leaves won't turn the ground too acidic. I like a soil pH for red raspberries between 6 and 6.5.

I try to pile the mulch in the fall *between* the rows, allowing as few leaves as possible to snuggle up next to the plants. I want the ground around the plants to freeze over wintertime. That's just a personal foible perhaps, but most fruits seem to need to go through a period of freezing weather during dormancy to continue in vigor. In spring, after the ground warms up, I spread the mulch up among the canes, effectively stopping most weed growth.

On the subject of winter freezes, I should also add that too much cold weather is just as bad as not any—if not worse. Red raspberries are generally quite hardy—hardier than blacks—

but they will winterkill occasionally. The berry kills back more where winter temperatures seesaw above and below freezing regularly than where cold weather comes and stays until spring.

To prevent winter damage, or at least to give your plants a better advantage, do not fertilize them in summer. That encourages lush growth late in the year—the kind of growth that winterkills most often. Mulching can cause lush late growth too, if August and September are very wet. Ideally, mulch should be applied in June and removed in late August on raspberries—with a good cultivation just before the mulch is put down and just after it is removed. But that increases your work too much and results in much less mulch rotting into the soil (and therefore much less fertilizing value) than a permanent mulch.

A good coating of manure every 3 years (or whatever application of organic fertilizers you use) should suffice to keep your plants in reasonably good production if you attend to your pruning assiduously. Apply rock phosphate every 4 years.

Unless you have somehow managed to avoid infection of virus diseases or live where the diseases are not prevalent, your raspberry patch will start decreasing in production after about 7 years. The decline in production seems inevitable after 7–10 years, and it is usually linked to virus diseases, but maybe it's just old age. At any rate, you will probably have to get a new planting started every 7 years or so. When the new one is producing well, tear the old one up and use the land for vegetables or grains. Do not return to raspberries for at least 5 years.

Reds and Yellows, Purples and Blacks

Managing everbearing red raspberries is somewhat different than regular red varieties. First of all, "everbearing" is a misnomer. So is "fall-bearing" in a way. These types of red raspberries should be called "two-season" berries. They produce a crop in early summer and a crop in fall.

Understand their growth habits. The canes that fruit in summer will then die, as with regular red raspberries. The new canes that come up in the spring, however, fruit in the fall and again the following summer.

If you want to manage the two-season varieties to get both a summer and a fall crop, then you prune as you would for regular varieties. You will get a small crop in the fall and a small crop the following summer.

I manage my two-season varieties so that they produce only a fall crop. This is the easiest way of all to handle raspberries. After the fall crop is finished by hard frost, I mow down the canes to ground level. If I suspect disease, I remove the mowed and shredded canes; if not, they remain as mulch. Anyway, that's all the pruning that is necessary. Next spring, new canes come up which produce a fall crop. Without the old canes there to produce a summer crop, the fall-bearing canes have no competition for water or nutrients and therefore produce a more bountiful fall crop than they would otherwise. And as everybody knows, fall red raspberries command very high prices on the market. So high in fact that you can feel very rich while you are eating yours for breakfast every morning in September.

Speaking of September, that's the name of one of the better two-season varieties. Another I like is fairly new, Heritage. Southland, already mentioned, is another new two-season type that is adapted to the mid-South. Scepter is recommended for the mid-Atlantic region. Some older varieties I've had, like Fallred, grew vigorously, but the berries tended to be crumbly, falling apart when picked. My present feeling is that Heritage is the best two-season variety for the East. September is okay, but in my garden, the canes aren't quite vigorous enough to suit me and the berry is a little on the firm side —not of the best eating quality. Heritage doesn't fall over as badly as other varieties, and its fruit tastes better.

I tie my fall-bearing canes to a single overhead wire about 4½ feet off the ground. August is the time to do the job. Without tying, the fall-bearing canes fall flat on the ground, usually allowing too many berries to rot as they ripen.

Right now, I'm most enthusiastic about my yellow raspberries, which I grow the same way as my regular reds. The name is Amber, which presently is the only yellow variety offered by nurseries I buy from. The first 2 years my Ambers were growing, I was quite disappointed. They just sort of sat there and scowled back at me, making poor growth and poorer berries.

But then they took hold. New canes were more vigorous and, for the first time with *any* raspberry in my garden, were strong enough to stand upright without staking or trellising. Moreover, berries that had been runty and crumbly now grew big and juicy. The canes do not produce heavy yields, but the berries that are produced are of good eating-quality. The taste is sweeter than of any other raspberry; honey-

The yellow raspberry is an especially desirable mutation of the red variety. It is quite rare.

colored, the berries seem honey-dipped. An application of wood ashes seemed to help them gain in vigor, which leads me to believe that yellow raspberries like a little more lime than other varieties.

Black and purple raspberry varieties have a growth habit unlike reds and yellows, and so they must be handled differently. The blacks and purples fruit once a year, in midsummer, on last year's canes, the same as red raspberries. But the plants do not increase by suckering. Instead, canes in late summer, having grown to a height of 6 or 7 feet if not pruned, gradually bend over to the ground. The tip of the cane pushes into the soil and roots, thus forming another plant. Clip it free from the cane the next spring, and you can transplant the new plant to start another row. Or simply by manually directing the canes so they root where you want them to, you can start a new row next to the old one, or fill in between plants in the row.

However, once you have your rows established and do not need more plants, you must prune the canes to keep them from tipping over and rooting. When new canes reach a height of about 5 feet (before they bend over toward the ground), pinch their tips off. Usually you'll be doing this in late July and early August. When you tip prune a black or purple raspberry cane, you force the plant to branch more and increase the stoutness of the cane to the point where it may stand upright without support of a trellis.

In addition to late summer tip pruning, you should, after the crop of berries has been harvested, cut out the old bearing canes, just as you do with red raspberries. Tip pruning may encourage too much side branch growth, which should be trimmed back in spring before growth starts. But there is no further winter pruning necessary on blacks and purples, unless canes die back from winterkill.

I train my black raspberries (often referred to as blackcaps) to a 2-wire trellis, the 2 wires parallel to each other about 4–5 feet from the ground. The wires are attached to the ends of crosspieces nailed to posts set at the ends of the row. Between posts, stakes on either side of the row every 15 feet further anchor the trellis wires.

Fruiting canes are tied to the wire on one side of the row. After they bear, I cut them out and at the same time, I tip prune and tie the new canes that will bear next year to the wire on the *other* side of the row. This trellising system keeps bearing canes more or less separated from non-bearing ones. Picking is much more convenient.

A somewhat easier and more commercial way to trellis blackcaps is with a single wire directly above the plants. Plants are severely pruned to maintain a narrow row that is easy to keep free of weeds.

Plants are tied to the overhead wire, and not allowed to grow much more than 5 feet tall. Bearing and non-bearing canes alternate—very neat and making the berries easy to pick.

Because black and purple raspberries are very brambly and will spread quickly into a miniature jungle if you let them, I advise almost ruthless pruning. Keep the row narrow, no more than a foot wide. Allow no more than 1 cane every 6 inches. The fewer canes per plant, the bigger the berries.

With blackcaps, that last observation is important. Blackcaps tend to be seedy anyway, and small berries are almost too seedy. Much better a single plump blackcap than 4 small seedy ones.

Purple raspberries, a cross between reds and blacks, are juicy and soft, no matter what size. In this regard, they more closely follow the reds. In taste, they are tarter than either reds or blacks. They make good jam. Purple raspberries have not enjoyed the popularity that reds and blacks have, but I don't know why. I like them better than reds or blacks. Clyde is the variety now most in favor in New York, where purple raspberries are still grown commercially. In Michigan I have seen some splendid plantings of Sodus, an older purple variety. Another purple, Amethyst, has been developed in Iowa and has received praise from plant breeders all over the Midwest.

In black raspberries, Cumberland is still the old standby throughout the Midwest and East. Cumberland seems to be less susceptible to winterkill than some of the newer and better quality blacks. Plum Farmer is another old and acceptable variety for the Midwest. Newer black varieties grown especially in the Finger Lakes region of New York are Allen, Bristol, Dundee and Huron. Although no black raspberry grows well in poorly drained soil, Dundee will tolerate tighter soils than Bristol. Huron is a late season berry somewhat resistant to anthracnose. Logan is another variety available from eastern nurseries. It's an early variety, ripening about a week ahead of Cumberland. My experience is limited to Cumberland, Bristol and Allen. The latter two seemed more susceptible to winterkill in our fluctuating winter temperatures, but otherwise, I haven't seen much difference among the three. Logan is also adapted to the Rocky Mountain area but will need some winter protection, as will any other black variety grown there. On the northern Plains, no black raspberry is recommended. Too cold. Morrison, in addition to Dundee and Plum Farmer, is grown in the Northwest. Black Hawk is excellent for the Midwest, from the Great Lakes through Iowa, where it originated.

Besides Amber, the only other yellow cultivated raspberry I know of is a variety called Goldenwest, developed for the Pacific Northwest. There are probably other varieties. Amber is available, if your regular nursery doesn't have it, from the New York State Fruit Testing Cooperative Association, Geneva, New York 14456.

Red and yellow raspberries will stand colder weather better than blacks. I should mention too that while I tip prune my blacks in late summer, some growers do not agree with this practice. They think tip pruning encourages late growth that is more susceptible to winterkill. Where black raspberries grow with great vigor and where virus is absent, canes can be allowed to keep a length of 8 feet or more and can be wrapped around a trellis rather than cut back and tied.

The Fruits of Your Labors

What kind of production can you expect from raspberries? Figure between 4,000 and 8,000 pints per acre. That's a lot of leeway, but it gives you a ballpark idea. Good producers can do better than 8,000 pints too. If you can sell your berries for 60 cents a pint, you can quickly understand the profit possibilities of raspberries—when you can get them picked!

Picking raspberries is slower than strawberries because the berries are smaller, though you don't have to stoop over so far. Great care must be taken with raspberries. If not quite ripe, they will come from the stem with difficulty—the picker will usually squeeze too hard and injure the berry.

Never pick into anything larger than a pint box. Better a pint container that is broad and shallow, so that you pile no more than 3 layers in the box. Raspberries are so fragile that if you pile them much higher than that, they will mash together too much.

When the berries are ripe enough to slide easily from the stem, get them picked quickly if you are trying to make money by selling them. If rain falls *hard* on ripe berries, it can ruin them quickly.

If you are serious about selling raspberries, put some kind of supplemental irrigation in your plans. You may not need it every year, especially if you mulch continuously. But particularly a fall crop may run head on into a drouth when you're counting on profitable sales.

Spray irrigation is often used on raspberries, but it is not the best method. Water spraying on nearly ripe berries does them no good, and it is better to use some kind of drip or trickle irrigation system. Perforated plastic hose is now on the market, from which water will trickle gradually. A long length of this hose stretched along the base of the plants under the mulch will provide sufficient moisture while using much less water than spraying.

Freezing raspberries will pay handsomely. At least we think frozen raspberries make good eating in the off season. But only if you freeze them correctly. Don't try to wash them. If you raise the berries without poisonous sprays, on mulch, there is no need to wash them anyway. Pick out nice berries and put a layer of them—just one layer—on wax paper in a tray. Set the tray in the freezer. The berries will freeze hard as marbles and you can then put them into pint or quart boxes for freezer storage. This method prevents the berries from becoming a mass of goo when thawed for eating. Each berry retains its texture.

Foraging for Raspberries

If you don't have time or space to grow your own raspberries, you can forage for wild ones. Reds don't grow wild, but yellow ones do. I've found them in Ohio and Michigan, growing in the same general areas as the much more plentiful wild blackcaps. The latter grow through the Midwest and Northeast and, for all I know, in the West too. Look for them along brushy creek and river banks, in abandoned pastures, cutover forest land, second-growth woodlots, fence rows, and along country roads.

Along country roads. Not so many years ago this was one place everybody could walk without trespassing to gather wild food or flowers or simply to watch birds and enjoy nature. The roadsides were grassy down to the ditch, then bushes of all kinds grew between ditch and farmer's fence

Black raspberries ripening. They have the same nutrients as the red raspberry, lacking only the vitamin A found in the latter.

resulting in a kind of hedgerow. Now every rural township seems bent on killing roadside "wild" areas with herbicides. Then the counties turn around and want to spend a big wad of money to establish "nature" areas for people to hike through.

"Wild" roadside fence rows do not interfere with farming that much, nor do they take very much land out of production. Maintaining hedgerows in such places is not only beneficial to man and nature alike, but provides jobs for "unskilled" road workers who have trouble enough finding employment today.

The country road past our home farm was a bread basket of wild foods and natural delights when I was younger. Not only raspberries, but strawberries, blackberries, elderberries and wild grapes grew there. There is no smell so pleasantly aromatic as wild grapes in blossom—none, anyplace. Wild asparagus, poke and dandelions were available too. Celebrations of the arrival of spring beauties, buttercups, black-eyed Susans and wild roses along the road were spring and summer rituals. And gathering bittersweet in the fall.

And the birds. Larks and bobwhites on the fence that divided road from field. Red-headed woodpeckers gliding from one telephone pole to the next. Mourning doves on the telephone wires. A high-hole (which you probably know as flicker) with a nest in the hollow of a tree nobody had bothered to cut out of the fence row. Bluebirds nesting in the wooden cornerposts of the fence. A pheasant sitting on 22 eggs, deep in the grassy cover of the ditch.

I could have shown you all that in a mile walk along our country road in 1950. But not today. There is no brushy fence row there anymore. No fence at all. When weed killers came along, our family got the spraying fever. And rejoiced. No more cutting weeds on hot August afternoons. Just spray away the work.

The road department decided to spray away their work too. They had more money to spend on herbicides than we did, and more time for spraying. They sprayed hell out of everything they could reach from the road. They killed the elderberries, the blackberries, the raspberries, the wild grapes, the bittersweet and most of the flowers. The road workers sprayed trees and other ornamentals planted in front of houses. Their spray drifted into gardens and lawns, knocking out flowers and grape arbors. Soybean fields unfortunate enough to lie alongside roads were killed back or badly wilted 10 feet into the field.

But by heaven, we now have clean fence rows. Nice neat grass, with no blemish of brush between road and field. The next step, I suppose, is to pave the whole blasted roadside with green cement. After that job is finished, the road department can fire all the unskilled help who can then go on welfare. And the public can go to costly parks on Sunday where the signs say: do not pick the bittersweet; do not pick the wild flowers; do not. . . .

But what do you say—to someone who has not smelled the wild grape in bloom? Or picked a box of wild raspberries on a dewy morning?

Raspberry Ailments

There's a number of diseases and bugs that can make life

miserable for your raspberries. In a garden, if not a larger commercial planting, proper organic methods will generally keep most pests under control. Those diseases that organic methods won't control can't be stopped by chemicals either.

"*Virus decline*" is the most serious threat to raspberries. In actuality the term covers a number of virus diseases—mosaic, leaf curl, streak and some viruses that have no apparent symptoms other than a general decline in vigor and production. Scientists can now detect virus (and therefore produce virus-free plants) by "indexing." They graft a leaf from a raspberry plant to another kind of plant which is especially sensitive to viruses. If the virus is present in the plant from which the grafted leaf came, the indicator plant will show distinct symptoms even if the symptoms did not appear in the raspberry plant.

Mosaic, the most common virus in raspberries, is very visible. Leaves turn a mottled, yellowish-brownish-purplish color. Sometimes leaves become puffy or warty. "Leaf curl" results in small leaves that curl downward. In "streak," bluish streaks appear on *new* canes. The most important point to remember is that symptoms of virus diseases can easily be confused with symptoms of nutritional deficiencies or of other diseases. Even experts may have a difficult time diagnosing the trouble. So don't hit the panic button every time a raspberry leaf turns yellow. It might be nothing more than poor drainage. Only if your plants continue to decline, after you have taken care to correct for other deficiencies, should you conclude virus disease is at your plant.

There is no cure for virus diseases in raspberries other than heat treatment, which can be done only in a laboratory. Rather than cure, you must take every possible step to prevent virus. Buy virus-free plants. If you can, establish your patch or field at least 500 feet (1,000 is better) from wild berries or old tame ones that might possibly be infected. Some varieties are resistant to some viruses, or seem to be. Among supposedly resistant varieties are the reds: Taylor, Viking, Washington, Milton, Indian Summer and Latham. However, that does not mean these varieties are resistant to all viruses. Latham and Taylor, if not the others, can be purchased as virus-free indexed plants, and I strongly advise you to do so.

Aphids carry virus diseases from infected to clean plants. Spraying insecticides can give only partial protection because a virus-carrying aphid can infect a plant before the spray knocks him out. Because aluminum foil, used as a mulch, seems to deter aphids from attacking vegetables, scientists have suggested using it around raspberries. The aphids might possibly stay away, but I don't know of any grower who has tried it on raspberries.

Once a plant is infected, all of its parts contain the disease. You can't take suckers or tip plants that look healthy and transplant them someplace else to escape the virus. The disease will show up in the new plant if the mother plant was infected. You may, however, get fair production from the transplant for a year or two.

Anthracnose, a blight disease, is the second most serious raspberry ailment. Its other name, gray bark, gives you an indication of how to identify it. First, reddish-brown or purplish spots appear in new shoots. The spots get larger, grayish in the center, purple around the edges. In severe cases, the canes get a gray film or crust on them and usually winterkill. Black raspberries are more susceptible than reds.

As with virus, sanitary methods are about the only way to control anthracnose. Cut out old canes as soon as they are finished bearing. Keep the row clean of weeds and excessive cane growth, as anthracnose has a better chance of flourishing where air passage is poor. Some varieties show a little more resistance to the disease than other varieties. Among the blacks that includes Blackhawk, Dundee, Evans and Quillen;

purples: Marion and Sodus; reds: Cuthbert, Indian Summer, Latham, Newburgh, Ohta, Ranere and Turner. But rather than try to choose a resistant variety, it is better to buy good certified disease-free plants from reputable nurseries, then to follow good sanitation practices.

Lime-sulfur is the old standby spray administered in early spring to ward off anthracnose. Sulfur is a no-no for strict organicists, though I have argued (unsuccessfully) that sulfur is a naturally occurring mineral and ought to be approved for organic use. But at least for now, you can't use sulfur if you intend to sell certified organic produce.

Crown gall can sometimes be serious in raspberries. This disease is caused by a bacteria rather than a virus, and there is no cure for it, chemically or otherwise, once a plant catches it. Rough, warty galls appear on the roots, the canes become stunted, berries seedy, and the plant finally dies. You *can* sterilize the soil, and you can also avoid replanting for 3 years where crown gall has occurred. Your best preventive is buying certified disease-free plants from good nurseries.

Spur blight makes purplish spots on the lower parts of new canes, usually around the spurs (latent buds from which branches can grow). The disease is not serious unless heavy rank weed or cane growth combines with unusually wet weather to cause a buildup of the fungus that causes it. If that happens, the spots can spread completely around the cane, seriously affecting its growth and even killing it. Keep the row open for good air passage, and normally spur blight won't become a serious danger.

Verticillium wilt, the old bugaboo of vegetables like tomatoes, potatoes, eggplant and peppers, will sometimes attack raspberries, just like it will strawberries. There's not much you can do about it either. Even chemical users must just hope and keep their fingers crossed, because a spray won't help. The wilt enters raspberries from the soil through the roots.

Verticillium hits healthy young canes, causing them to wilt suddenly. Bad infections might show up with a bluish stripe running from the base of the cane upwards. About the only precaution you can take is to not plant raspberries where the above mentioned vegetables have shown symptoms of verticillium wilt.

Diagnosing wilts and blights correctly is no easy matter, as I have said before. I'm sure that most of the experts will agree that there is much more to be discovered about wilts like verticillium. I've had enough experience to know that wilts do not seem to have any rhyme, reason or consistency about the way they appear and disappear. One year, some of my red raspberry canes suddenly wilted and died in August. Surely verticillium wilt, I thought, since there were no signs to suspect anything else.

Maybe it was, but healthy plants eventually filled in the space where the strange wilting occurred, and never has it returned. The plants have remained healthy ever since. I even performed the final perfidy of growing potatoes and tomatoes between the rows of these same raspberries, with no apparent ill effects. (I don't recommend doing that! If you want to make use of the space between raspberry rows by growing some non-competing vegetable there, try members of the cole family—cabbage, broccoli, cauliflower.)

I have also observed how muskmelon and cucumber vines will suddenly wilt and die while a plant of the same variety nearby (and sometimes another vine of the same plant!) does not wilt. Look for borers when that happens. But wilts have never caused crop failure in my garden, so I don't worry.

A few other blights, mildews and rusts may attack your plants, but none of them is serious except on rare occasion. The preventive sanitary precautions already advised will generally control them. Fruit rots may also be troublesome in extremely wet years but they are not worth worrying about.

Insect Pests

Only 2 or 3 bugs represent serious threats to your dreams of raspberries and cream. And in general, not even these bugs present too much of a problem for organicists.

The raspberry cane borer will hollow out a cane now and then. The female beetle deposits an egg toward the top of the cane. When the egg hatches, the larva eats its way down through the pith of the cane. The cane eventually dies.

When the tip of a cane wilts over during the summer, there's probably a borer inside. If you look closely you may see, right below the wilted portion of the cane, two girdling marks that the beetle made and between which she deposited the egg. Usually if you cut the tip off below the lower girdle, you get the young larva. Burn the tip. You can usually control the borer with this kind of hand control.

Sometimes in extremely hot weather, cane tips will wilt from the heat. The rule is: Don't pinch off until you know what you're pinching.

The red-necked cane borer can be a nuisance too. Just this morning I noticed 2 of my yellow raspberry canes dying as a result of this pest. Had I been more vigilant, I could have saved them.

Usually you notice this borer first by cigar-shaped swellings or galls near the base of a cane. If you cut into them soon enough, you can find the borer and kill it. Only the borer's tunnels are left in my infected canes. But I will trim away infected parts and burn them anyway. Even if the wilting canes do not die completely, they would almost certainly winterkill later on.

The red-necked borer is often found on wild blackberries and wild roses. If you have any of these nearby, it is best to get rid of them.

Get into the habit of carrying a pocketknife with you as you make your daily rounds of the gardens. You are always ready then to make minor operations on suspicious swellings in plants, or to do a little necessary pruning. When you do not have a knife along, you have to postpone such jobs, usually promising yourself that you'll "do that tomorrow." "Tomorrow" you are in just as big a hurry as you are today, and you take a different route through your homestead. You don't see the stem swelling this time to remind you. Bug control measures that could have been used conveniently a little at a time thus accumulate for a heavy Saturday. And in the meantime, the bugs have been chomping away.

The raspberry crown borer is another of the nefarious family of tunnel-makers that gardeners could do so well without. This fellow likes red raspberries best and seldom attacks blacks. The larvae move down the canes and burrow into them at about ground level. You can rarely spot evidence of them once they are into the crown, but the plant makes poor growth and any cane that doesn't amount to much by late summer ought to be pruned out anyway. Pull up an infected cane and split the crown with a knife. If you find a white larva in the pitch just above ground level, that's a crown borer.

The eggs of the borer are laid on the underside of the leaves in the fall by a clear-winged moth, say the experts. I've never found any, but if you cruise your patch in September looking for egg masses, you should be able to effectively control the pest in a small planting. As far as I know, I have not been bothered by this borer—but I'm knocking on wood.

At any rate, if you clean out old canes after they have produced a crop of berries and paid attention to all other sanitary precautions I've mentioned, you should not have too much trouble with borers.

The same is true of the European corn-borer, which sometimes likes to vary its diet with a red raspberry cane. The cane will usually break off in late summer if infected, just like a corn stalk will. Corn borers—this type—are a little over an inch long, and greenish-brown.

If you live where the Japanese beetle roams, expect this voracious devil to take a liking to your raspberry leaves. I can't say the shiny green bugs have harmed my patch, but they are most injurious to the patience I try so hard to cultivate. Seeing a band of Japanese beetles larking around my raspberries, reducing leaves to mere webbings of vein, brings out the savage in me. In the evening when the glutted beetles are too tired to do much flying, I move through the garden, brushing beetles into a jar or can of gasoline. One pass over the grapes and raspberries and roses will usually fill a tomato can. And I enjoy the massacre.

In hot dry summers, you may notice your raspberry leaves looking pale and dull. It may be more than just dry weather. Check the underside of leaves with a pocket magnifying glass (something else handy to carry with you in the garden). You may find spider mites. The best insecticide for controlling them is water. A well-watered raspberry growing vigorously will seldom be harmed much by spider mites. A hard spray of water, some say, will knock spider mites off some plants, such as evergreens.

Sometimes little fruitworms get into the berries, but I doubt if this will ever be a problem if you pick your berries as they ripen. And I'm sure you'll do that. If you don't, someone else in the family will. The berries taste too good to resist, and that can result in another "pest" problem. I let my last picking go a day too long, and my son, with eager help from some of his friends over playing football, was only too glad to "help" me out. "That's okay, Dad. Some more will be ripe tomorrow." Yea.

Chapter Four

True-Blue Berry Lovers Love Blueberries

Twenty years ago, a horticultural writer might have hesitated to include blueberries in a berry book intended for a national audience. But the relatively infant blueberry industry in the United States has come along fast and strong with better varieties and improved cultural methods to the point where some kind of blueberry should grow wherever soil acidity can be maintained in a pH range of 4 to 5.5. As a matter of fact, if one includes all the many wild subspecies of blueberry along with domesticated varieties, blueberries can now be found in parts of just about every state east of the Mississippi, plus a number of states immediately west of that river from Texas to Minnesota, plus the Pacific Northwest. Only in the alkaline soils of the West and Southwest is the blueberry family absent over large areas.

The many colloquial names for blueberries show the diversity of the family (referred to as *Vaccinium* among the experts): whortleberry, hurtleberry, whinberry, blaeberry, trackleberry—not to mention related fruit like bilberry, dangleberry, tangleberry and huckleberry, which I'll talk about in another chapter. All these berries were known and cherished by the American Indian. He dried the fruit, in which condition it would last a surprisingly long time. I have seen dried blueberries that had to be at least 200 years old in clay pots dug from Indian burial grounds in Pennsylvania. Though of course no longer edible, the berries were in remarkably good shape.

Blueberry Culture

Six kinds of blueberries have commercial importance in the United States. 1). The highbush blueberry is the kind you almost always buy fresh in supermarkets. 2). The lowbush blueberry, which grows "wild," is managed for commercial production particularly in Maine. Lowbush berries are usually sold as processed rather than as fresh fruit. 3). The rabbiteye blueberry was domesticated from wild species in the Southeast. 4). The evergreen blueberry grows wild in the Northwest Pacific coast area and is sometimes planted as an ornamental. 5). The mountain blueberry is another wild species of the Northwest. 6). The dryland blueberry is native to the Appalachian mountain range from Georgia up to Pennsylvania and over in the Ozarks too.

Despite its ubiquitousness, the blueberry is one native fruit many millions of Americans have not yet eaten fresh. Mostly the reason for that is that there are relatively few areas where blueberries are raised commercially in quantity, and all these areas are on the edges of the country, far from where many of us live. Most blueberries on the market come from New Jersey, North Carolina, Michigan, Maine and Washington.

Until I was 34 years old, I had never seen a blueberry growing. I had done my living in various parts of the cornbelt where blueberries don't grow because the soil tends to be neutral rather than acid. Imagine my excitement then, when exploring the woodlot behind our new home in Pennsylvania, I peered out through the brush into a neighbor's small or-

Under a thin film of wild yeast, the blueberry has a shiny, dark skin.

chard to see shiny green bushes fairly dripping with misty blue berries. Anytime I find a new plant, especially something good to eat, it's a Very Important Event. The thrill of finding what I knew had to be blueberries was quickly followed by the pleasant realization that I would be able to grow them on my place.

And grow they did. I had always pictured blueberries as adapted only to sandy, desolate areas along seacoasts where nothing else would grow. And while blueberries will grow in many such places, they will do just as well, if not better, in rich garden soils, even clay soils, if such ground is acid enough.

The blueberry is our newest domesticated fruit. Only about 50 years have elapsed since Elizabeth White and F.V. Colville began to crossbreed selected wild blueberries from the Pine Barrens of New Jersey to develop the first highbush berries for commercial production. Since then, blueberry culture has grown by leaps and bounds—propelled onward by great consumer acceptance. In fact, no one knows when, or at what point, demand will peak and level off. Fruit marketers know, for instance, about how many apples can be sold in any given year in America. Not so with blueberries. Not enough people have eaten blueberries yet to know the ultimate level of acceptance. In all likelihood, consumption will increase steadily for some time. Not only does the berry taste good, it packs, ships and keeps well too.

In discussing blueberry culture, I'll treat the highbush blueberry first (because that's the one I have experience in growing) and then go on to the other types.

Growing Conditions

Can *you* grow highbush blueberries? If holly, azaleas and rhododendron grow well in your neighborhood, so will blueberries. For winter hardiness, any climate that peaches tolerate will be right for blueberries. (Lowbush blueberries can endure very cold temperatures when and where snow is deep enough to cover them in winter.)

Highbush blueberries like sandy soil if it has organic matter in it. That's why they do so well on the Pine Barrens of New Jersey. That soil looks like pure sand, but actually much of it contains 5% to 15% organic matter in the form of peat. The soil is well-drained, another requisite for blueberry success, but the water level is fairly close to the surface, lessening the danger of drouth.

These conditions are all supposed to be ideal for blueberries. However, I found I could raise good blueberries on clay ground with heavy mulching, so long as I had suitably acid soil. A pH of around 4.8 is ideal for blueberries, but mine, at about 5.2, was still okay. By adding oak leaves and peat moss, as mulch, I am able to increase the acidity a little bit.

If your soil is above 5.5 or thereabouts, but not much above a pH of 6, you might raise blueberries by artificially acidifying your soil. Sulfur alone will increase acidity, but organicists add tannic acid to their soil or patiently build up acidity with yearly additions of oak leaves, oak sawdust or oak bark, all of which contain tannic acid. Sumac leaves are very high in tannic acid and will make an excellent soil amendment for this purpose if you can find enough of them. Oak galls—those round brown balls you find on oak trees—are also very high in tannic acid.

But where natural acidity is missing from the soil, it will usually be difficult to make blueberries grow properly. If pH is over 6.5, I wouldn't recommend even trying to raise soil acidity artificially. However, if you want just a few bushes, you might try 1 or 2 alternatives.

The first is to dig a hole about 4 feet in diameter and a foot deep and fill it with a homemade acid soil. The formula for such soil: 3 parts sand, 3 parts sawdust or horticultural peat and 2 parts acid leaf mold. Make enough to fill the hole and plant your bush in it.

If the surrounding soil has been limed within the last 2 years or if it is naturally limey, your blueberry plant still won't grow. If that's the case, and you *really* want your own blueberries, you might try growing a few bushes in tubs filled with your homemade soil. Cut a 55-gallon steel barrel in half to make 2 suitable tubs. Set them in the ground so the rims are a couple of inches above the surrounding surface to keep out run-off water that might contain lime. Fill with homemade dirt as described above and a blueberry plant.

Instead of making a homemade "potting" soil for your blueberry bush, you may be able to find naturally acid soil in woodland—even in areas where the cultivated soil is not acid. Woods soil from forest land that has always been forest land, especially in oak or pine groves, should be fairly acid. To test for soil acidity, use a soil testing kit and follow directions. You can have your soil analysed for all chemicals and nutrients by sending soil samples to your state agricultural college. Contact the Extension Service office in your county for the proper procedure. Farm service stores often have soil sampling programs for customers too.

If your soil is extremely acid—below 4—you can add a little lime to raise pH to the optimum level for blueberries. It is best to do such liming the year before you set out plants.

When your soil pH is close to the upper ranges for blueberries, you must be very careful not to get any lime on the blueberry ground. A couple of years ago, I applied lime to some plantings next to the blueberries but on a higher elevation. Rain washed the lime down to the berries, and the next year the bushes looked as if they had caught yellow jaundice. One bush died and several others are pale and chloritic and may die yet. (Incidentally, you can hurt azaleas and rhododendron the same way, by liming lawn grass next to them.) You learn by living.

It goes without saying that you should not plant blueberries on poorly drained land. If your land is poorly drained because of poor percolation from tight clay soils, organic methods will eventually improve the tilth and solve this problem. If poor drainage results because the ground is lower than the surrounding terrain, the condition may be corrected with tile drainage. But such a low spot, even if drained, is not a good place for berries. In cold weather, it will always be colder in the lower spots where air drainage is poor. Frost and freezing is always more likely there.

Secrets of Gardening Success

Mulch is doubly beneficial on blueberry bushes because the plants do not root deeply and therefore are easily harmed by drouth. The heavy coating of oak leaves I put on my blueberry ground has meant that I have never had to water them. Other good mulches are peat, sawdust, pine needles and, if you happen to be located near a brewery, spent hops. As with raspberries, I try to keep the leaf mulch (which I apply in late fall) piled in the middle of the space between the rows over winter, leaving the ground at the base of the plants uncovered. Blueberries need a chill period during dormancy in order to grow vigorously the next year. Then in the spring, I spread the leaves out evenly.

Mulch becomes almost a necessary cultural practice on highbush blueberries grown in the south—not only to save moisture, but to hold down soil temperature. Rotted sawdust is an excellent mulch for this purpose.

I study my blueberry mulch with keen interest as it rots into compost and becomes a part of the earth. I've not cultivated the blueberry ground ever, so the soil layers are not disturbed much. After 7 years there's about a 4-inch layer between original soil and the top mulch so rich it lifts your spirit just to dig your fingers into it. What's more, this layer has become a reliable fishworm bed whenever we need bait for a fishing trip.

Just a word of caution. As I'll point out a little later, certain pests of blueberries are best controlled by cultivation. So while I believe in mulch, if I had a problem with such pests, I'd turn to cultivating temporarily.

The rotting mulch is all the fertilizer my blueberries normally get. I can't see any use in applying more fertilizer. The bushes bear so heavily now, I can't figure where more berries would find room. But where mulching is not practical or soil is poor, extra fertilizer will be needed. In 1921 Dr. Colville, the

Surrounded by a thick mantle of mulch, this blueberry plant is ready to meet the winter.

original Mr. Blueberry who developed the first commercial varieties, *tripled* yields of blueberries on poor sandy soil by using a mixture of sodium nitrate, dried blood, steamed bonemeal, phosphate rock and potash. He concluded, however, that if the soil contained enough peat, fertilizer was not needed.

Oddly enough, in the early years of blueberry culture, manure was thought to be detrimental to the berries. Experience has *not* borne that out, but neither has manure improved yields to any appreciable degree. If your plants do need more nutrition, use rotted leaves or sawdust and save your manure for other crops. Or take a tip from Colville and try dried blood, steamed bonemeal and phosphate rock.

Where blueberries are grown on what I call narrow-purpose soil (sandy, highly acid soil where few other crops can be grown), balanced fertility can be critical. Not only the balance between major nutrients—nitrogen, phosphorus and potassium—but between minor trace elements too. Of these latter elements or micronutrients, boron, magnesium, calcium, zinc, sulfur, iron or manganese might be deficient. Good soils fertilized with manure almost always produce or maintain proper micronutrient content naturally. But not with poor soils. Interestingly, in some experiments, blueberries grown on sandy soil deficient in boron, calcium and iron showed no symptoms of these deficiencies when the sand was mulched with peat. Moral of the story: Peat is a darn good mulch for blueberries.

What varieties of highbush blueberries should you grow? Above all, grow more than one kind. Blueberries are somewhat self-sterile and will yield better if 2 or more varieties are grown together. Also, consider planting an early, midseason and late variety to stretch the season out as long as possible. In the 3 chief commercial areas of highbush blueberry production these varieties (plus some new ones not yet generally available to home gardeners) are grown:

Michigan—*Early:* Earliblue; *Midseason:* Blueray, Bluecrop; *Late:* Jersey, Colville, Darrow.

New Jersey—*Early:* Earliblue, Blueray, Ivanhoe; *Midseason:* Bluecrop, Berkeley, Collins, Elizabeth; *Late:* Herbert, Darrow.

North Carolina—*Early:* Morrow, Angola; *Midseason:* Wolcott, Croatan; *Late:* Murphy, Scammell.

Some of the older varieties like Jersey, Rancocas, Weymouth and Stanley are fairly good varieties in the north or middle Atlantic states, though generally their berries are not as big as the varieties listed above.

Experts advise that you purchase 2-year-old plants to start your blueberry venture. Set them out in early spring, spreading the roots out carefully as you would when transplanting any other bush. Don't put any *un*rotted manure or sawdust in the hole. A little acid compost won't hurt, but it isn't necessary. Set the plant in the ground at the same level it was growing in the nursery. Mulch it and leave it alone for the summer.

The next spring, you can prune out any weak growth. If the plants seem a little peeked, you can cut off some of the fruiting wood. Watch that leaf growth is equal to fruit growth. Don't have the poor little bush all fruit and no leaves.

The third spring you'll get a couple of pints of berries. The yield will go right on up to 10 pints or more per bush by the time it is 8 years old, if you prune properly, and I'll get to that in a little bit. I don't know how long a blueberry bush can last. I know some good bearers over 25 years old.

Propagation

What do you do when you want to increase the number of bushes in your planting or to sell to other gardeners? Cuttings from blueberry plants will root fairly easily, and though it takes a couple more years to grow a cutting into a producing bush, you can do it without investing much cash.

Except for Bluecrop, most of the newer varieties root easily; of the older ones, Jersey, Rancocas and Stanley have reputations as easy rooters. In the early spring take cuttings from the past season's growth, from healthy shoots that have hardened off well but have a good spring to them. Each cutting should be about 4 inches long. It will root better if it contains only leaf buds, not any flower buds. The difference? Leaf buds are thin and flattish; flower buds are round and comparatively fat. The closer to the bottom the cutting is taken, the easier it will root. The top cut of the cutting should be made right above a leaf bud; the bottom cut right below a leaf bud. The sharper your pruning shears the better; bruising the outer layers of the shoot with a dull clippers deters the formation of roots at the cut.

Stick the cuttings in a rooting medium of half sand and half horticultural peat, allowing only the top bud to stick above ground. Then you must maintain a delicate balance of adequate moisture without overwatering. Commercial growers often use one of various automatic misters that spray a water mist on the cuttings at regular intervals, keeping them moist but not sopping. My cuttings have rooted well only when I cover the flats with clear plastic film, sealing moisture in and dry air out. For at least an hour every other day, I pull the plastic back so that air can circulate over the cuttings. If they are kept too wet, the danger of fungal diseases increases greatly.

Don't expect all your cuttings to root. With my more or less crude (cheap) methods, I'm happy if half of mine root. Don't disturb the cuttings that leaf out until the following spring. Then they can be planted in a nursery. Cuttings that die or are obviously sick should be pulled up and thrown away.

Cuttings can also be taken from this year's growing shoots. These are called softwood cuttings and are best cut from the bush just about the time the first fruit is bluing up. Small side branches from the shoots are cut, leaving a small spur of the main cane on the butt of the cutting. Cut the tip off so the cutting is about 5 inches long. Leave 2 leaves on the cutting; pinch the rest off. Then handle the same way you do a hardwood cutting. Chances are you won't have as much luck with softwood cuttings as with hardwood.

Mounding is another way to start new plants. Build a frame around the base of the blueberry bush about a foot high and 3 feet square. Fill the frame with peat and rotted sawdust, mounding the compost up around the canes of the bush. Roots will form on the sides of the canes under the peat-sawdust mound. They can then be cut off below the new roots and planted elsewhere.

I have used a third method to increase the number of bushes in my planting—in fact I may be the only one who has used this method. I simply dug up an old bush, divided it right down the middle and planted both parts. They grew as if nothing had happened to them.

That's what farmers meant years ago when the motto, "Grow two blades of grass where only one had grown before," was so popular.

Plant your blueberry bushes about 5 feet apart with rows at least 8 feet apart. That'll give you some 1,089 plants per acre, in case you're curious. The plants in a small patch may be spaced 6 feet by 6 feet if you need to conserve space. The width between rows on larger plantings will be dictated by the type and size of your cultivator.

For cultivation in blueberries, a disk is the best tool as it can be set so that it will not cut the roots, which grow fairly close to the surface around the plants. A normal cultivator of the shovel type is apt to dig too deeply and tear up roots. A disk, pulled behind a tractor, allows you to cultivate right up to the base of the bush too.

Birds, Bees and Blueberries

In small garden plantings, I heartily recommend that you screen your bushes to protect them from birds. Otherwise, you won't get many berries. Set posts 8 feet tall around the edge of your patch 8 feet apart and in rows through the patch 8 feet apart. Nail framing along the top of the posts, connecting 1 to the other in this fashion. Two-by-4s or even 2-by-3s make good framing material, though 1-by-6s are adequate too.

Next, nail wire screening (of a mesh small enough to keep out birds) all around the outside posts, as if you were putting up a fence. Be sure to allow a place for a gate of some kind. Then fasten screening over the top of the framing to form a "roof" over the bushes.

If you have not grown blueberries before, do spend a couple of years with them before you invest in a permanent frame and screening. If your bushes died from some nutritional difficulty or soil problem, you would no doubt want to move to another location to try again. Moving the posts and screens would just be additional work.

You can buy lightweight plastic screening or cheesecloth to drape over your bushes. Such screening affords some protection, but birds will still be able to get at many berries. It's much better to have a frame holding the screen at least a foot away from the bush. I know a fellow in Rhode Island who screened his blueberries with old fish netting. It sagged between posts, but worked fairly well.

I have had some luck luring birds away from blueberries with mulberries and other plantings. More on that in a future chapter.

Blueberries and honey go together in more ways than just good eating. Bees are required for good yields. Experiments by researchers at Rutgers University have shown dramatic increases in yield when bee population was kept high. In one case, bees caged on one bush so that they could not fly to any other increased the berry set on that bush by 7 times over surrounding bushes. A hive of bees per acre of blueberries is considered adequate for excellent pollination.

Naturally, an organic grower not using poison sprays will, or should, have better pollination from wild bees than other growers. But indications are that wild bees have been declining for other reasons besides spraying. Decline of bumblebees —better pollinators than honeybees because they will work in colder weather and are more vigorous—began before the advent of sprays, most entymologists believe. The decline has resulted from the destruction of so many of their natural nesting places—like the roadsides mentioned in the preceding chapter. Intensive cultivation and forest fires have taken their toll too. And how many of us humans are willing to endure bumblebee nests under the porches of our homes?

Since bees are so important to fruit growers, never miss a chance to encourage their protection and inform people about them. There are, of course, a few people who are seriously allergic to bee stings, and that's a different story. But bees present little danger to most of us. They do not sting unless forced to it. Bumblebees especially are very peaceloving until teased by boys. I say that after long experience at being a boy.

If in our outer suburbs, "wild areas" of brush and trees and fallen logs and ditch banks could be allowed to remain, we could conserve our wild pollinators quite easily. *We don't need to clear and mow into grassy lawn every inch of suburban land not covered by a house.*

Fortunately, such unkempt areas still abound in many suburbs. In our township, for instance, woodsy and brushy plots and lots are still fairly numerous. And enough of us are aware that nature is not always neat and prim, so we leave an unmowed corner "out back" whenever possible. Moreover, little spraying of insecticides is done here by the township. So, we have lots of bees—bumblebees, honeybees and some smaller kinds of insects that also help the pollinating process. If we get a good bloom, we get a good fruit set—no matter what the fruit.

Encourage beekeeping too. If you can't keep bees yourself, buy honey for your table instead of sugar. That way you help beekeepers make the profit they need to stay in business. You also help your own health. It's called voting with your pocketbook, a concept you can use as a powerful tool to make the world better.

Pruning

Pruning blueberries is not as simple as pruning raspberries. Watch how your blueberry bushes grow for a year or 2 (the first 2 years, very little pruning is necessary anyway) and you will begin to get a "feel" for what you must accomplish in pruning.

New shoots come up in early summer, grow vigorously, harden to a springy whip, maybe branch a little and develop fat buds by fall. From the buds on these "canes" there will develop the following summer clusters of berries. The canes will continue to branch in succeeding years with some new healthy growth. But as they branch this way and that for about 5 years, the ends or tips, where berries drooped in past years, tend to become what blueberry growers call "twiggy." This twiggy growth somewhat resembles elderberry stems after the elderberries have been removed. Twiggy branch ends continue to produce a few weak buds and therefore a few berries, but they need to be pruned off. Old (6 years) canes can be cut out to make way for the new shoots that come up each year from the roots.

The general idea is to keep the bush youthful and productive. Cut over-6-year-old canes at ground level; if a new shoot has grown from an old cane right above ground level, you can cut the old one above the shoot. It's better, though, to rely on new canes that come up directly from the roots.

If more thinning is needed, cut shorter branches off newer canes—try to select for cutting those branches that have the most twiggy growth on them. You may in the process cut out some good buds, but that won't hurt anything. The more a bush is pruned (up to a point, of course!) the bigger the berries will be. Most of us usually prune too little, not too much. There's some truth to the old saying that the best way to get your trees and bushes pruned enough is to get your neighbor to do it, while you do his. We all tend to go too easy on our own plants.

Prune out short soft shoots that grow from the roots late in the year. You'll be able to tell them from the good shoots. The former are pliable, but not springy, and usually a little flat on one side. Snip them out. It is through such soft shoots that a bush often contracts fungal diseases. Soft shoots usually winterkill anyway.

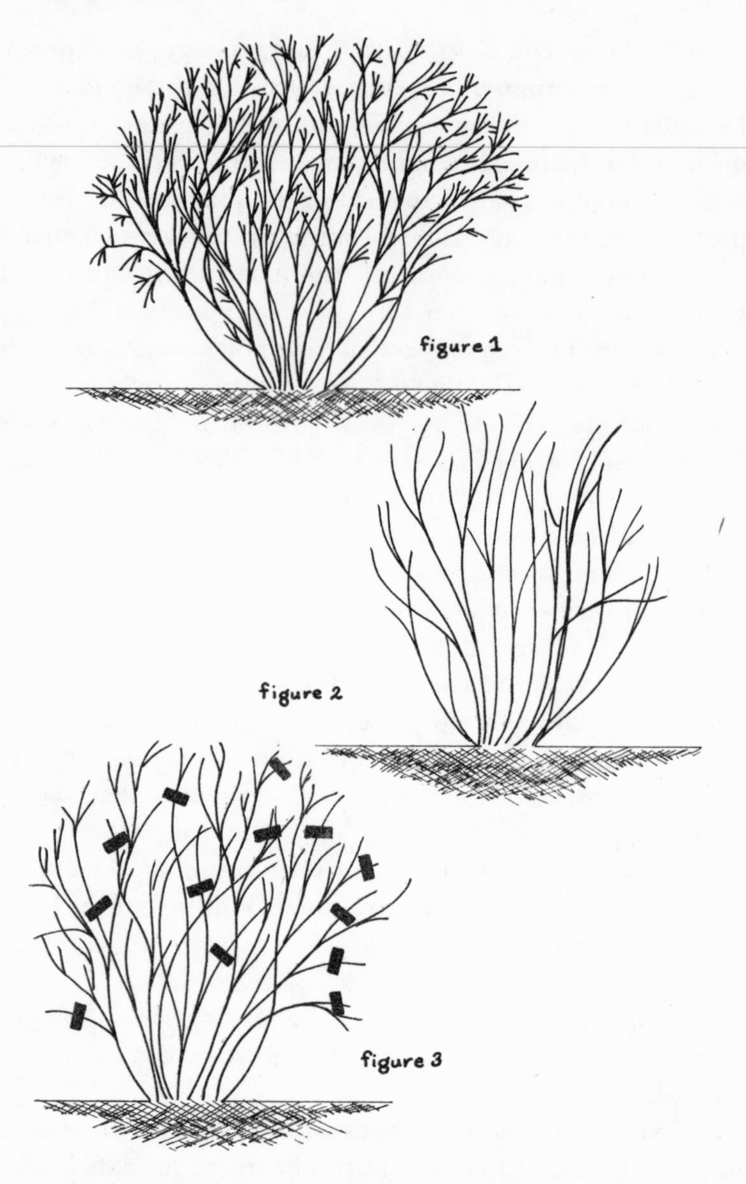

An old blueberry bush may be burdened with gnarled cane and twiggy growth, as in Figure 1. When properly thinned and pruned, as in Figure 2, will maintain vigor and production. Tip-pruning, suggested in Figure 3, can be beneficial too.

On the branched shoots you allow to remain, there will be some twiggy growth too. A skilled blueberry pruner will rub these stems and weak buds off with a gloved hand while his other hand is making other cuts with the pruning shears. You almost have to see someone doing it to grasp how it's done—I had to, anyway. About all I can say with mere words is to repeat: You want a few nice fat buds on clean, sparsely branched canes that taper from the thickness of your finger down to a pencil thickness; you do not want twiggy end growth full of small buds and dead stem ends.

Enemies of the Blueberry

Away from commercial concentrations of blueberry fields, gardeners and farmers with small plantings have not been bothered much by bugs and disease. Not yet, anyway. I don't know how long I will remain so fortunate, but in 8 years, neither bug nor disease has descended on my blueberry patch.

Nevertheless, be prepared, as the Boy Scouts say. The blueberry has one nasty enemy, the blueberry fruit fly, which has caused much trouble for commercial producers. The insect is the size of a housefly, with black wing bands that give it a somewhat zebra-like appearance. It lays eggs in the blueberries as they ripen. The eggs hatch into little colorless maggots in just a few days. You can hardly see them when they first hatch, even if you open a berry they infest. But in about a week, the berry begins to rot.

Control is difficult in native blueberry country, because the pest infests wild plants in the area. And wild huckleberries too. Before the advent of more "efficient" insecticides, standard control in commercial fields was a 2% rotenone dust, 25 pounds per acre. This worked pretty well (and still does) if you make 5 applications between June and the end of harvest. Because rotenone is harmless to humans it can be applied to berries for the fresh market right up to harvest time.

Two insect parasites attack blueberry maggots in the East: *Opius ferrugineus* Gahan and *Opius melleus* Gahan. Neither is too effective, however, even if protected by not spraying. But better some control than none at all.

Cultivation, too, helps control the blueberry maggot. Larvae that have been feeding on the berries drop to the ground under the bush, where they overwinter. Cultivating between the rows and hoeing between the plants exposes the larvae to predator ants and birds. Where the ground can be disked repeatedly, few pupating blueberry maggots will survive.

Another tip: early varieties are less bothered by blueberry maggot than late varieties.

The cherry fruitworm is another villain in the eyes of the commercial producers. In small plantings the worm seldom becomes a critical problem. It hatches usually in late May and feeds until the middle of June in New Jersey. After it has fed on a blueberry for a couple of days, it turns from a hardly noticeable, nondescript worm to a red one about a fourth of an inch long. The best organic and natural control is a parasitic fungus, *Beauvaria bassiana*, which attacks hibernating larvae of the cherry fruitworm and the cranberry fruitworm. The latter occasionally infests blueberries. According to Rutgers University bulletins, the parasitic fungus can almost cut the fruitworm population in half in wet falls and winters.

A very small wasp, *Trichogramma minutum* Riley, also parasitizes both kinds of fruitworms and is of "inestimable value" to the commercial blueberry grower, according to the book *Blueberry Culture*, edited by Norman Childers and Paul Eck, both of Rutgers. The wasp may keep fruitworm population at half of what it otherwise would be, but that still is not low enough to suit commercial growers, so they spray anyway. Spray kills more of the wasps—and the ants which prey on fruitworm pupae—which makes more spraying necessary. And the vicious circle grows more vicious.

An insect called the plum curculio will add blueberries to its repertoire of favorite fruits if given the chance. The adult curculio is a brown-snouted beetle about one-fourth inch long. Its mark is a small, crescent-shaped depression in newly formed fruit. The best spraying time is during the last part of the blossoming season, which would be a time especially harmful to pollinating bees. (Some commercial blueberry producers remove beehives from the fields during and right after spraying to avoid harming them.) Frequent cultivation is the best control organically, which means that if the curculio becomes a real pest for you, you may have to give up mulching for a while. Late ripening blueberries, especially Jersey it seems, are not much bothered by plum curculio.

The other member of the Big Four blueberry pests is the cranberry fruitworm, already mentioned. From eggs, a green caterpillar about three-eights of an inch long emerges and weaves a web around a cluster of berries. The web is fairly easy to spot and readily identifies the fruitworm. On small plantings, handpicking can keep the pest under control pretty well. Keeping weeds and trash cleaned up around the bushes helps, too, as does frequent disking of the ground between plants.

Scale insects might become a problem if left unattended (I've never had any on my bushes), but all the kinds that infest highbush blueberries are fairly easy to control organically. In fact two of the main pests, Putnam scale and terrapin scale, usually become serious only when pruning is neglected or when insecticides kill off the scales' natural enemies, such as the predator ladybird beetle. Again, that's what the book *Blueberry Culture*, mentioned above, says; I'm not just repeating a pet peeve. But you will have little scale trouble if you prune out dead and excess wood, use a delayed dormant oil spray (delayed meaning one of the very light miscible oils that can be sprayed right before blossoming without harming the new budding growth), and let the predators mop up any survivors later on.

Blueberry-bud mites are so small you can't see them with the naked eye. They can be a problem in the South, but rarely in the North. A miscible oil spray, as mentioned above, is recommended in late September or early October for control.

The blueberry-leaf miner eats a few leaves and may roll up a few more, but the bug rarely affects fruit yield. Perhaps for this good news we have to thank another parasite, *Apanteles ornigis* Weed, which researchers with no axes to grind call a "very effective biological control agent."

There are more insect enemies of the highbush blueberry. I doubt if the small organic grower will be troubled by them to any great extent, so why should I worry you by mentioning them?

On the subject of blueberry diseases, I must rely on what the experts say in their bulletins, because the only disease my blueberries have suffered from is "Too-Much-Lime-itis," as I have explained. There are very few direct cures for fungal, viral and bacterial diseases. The best advice I can give is to keep your pruning shears sharp and your fingers crossed.

Stunt virus, stem canker, mummy berry, botrytis blight and powdery mildew are the most serious of these diseases, and if that list of names doesn't scare you, there's also a witches-broom virus that attacks blueberries. Fortunately, the latter is not serious in this country. Fortunately again, the first 2 diseases mentioned are well on their way to being licked in cultivated highbush blueberries.

For any virus disease, control or eradication depends upon the same general methods used in strawberry and raspberry virus-free programs. Now that such programs effectively prevent infected blueberry plants from being sold, stunt is not the threat it used to be. However, the potential danger remains wherever wild blueberries grow. The chances of having stunt are greater in the South than the North. When stunted, the leaves of a bush are only about half-sized and cupped upwards. The plant may not die, but it won't grow either. Some chlorosis may be present.

I promised myself I would not use that word, chlorosis, in connection with blueberry disease, but there I did it. All chlorosis really means is that the leaves have paled from a healthy green to yellow. The yellowing could result from any number of causes. If someone tells you your plants are suffering from "chlorosis" he is telling you little more than if he says a sick person is suffering from paleness. Poor drainage will cause a plant to look chloritic; nutritional deficiencies (such as Too-Much-Lime-itis) will cause "chlorosis." In poorer soils, leaf yellowing may be due to a lack of iron, in which case the "disease" should be called iron chlorosis, if the word must be used.

Mummy berry pretty well describes the symptoms of this fungal disease. Berries shrivel and turn brown right before they should ripen. Leaves may wilt and blacken—looking as if a hard frost had hit them.

The fungus that causes this blight survives over winter on the ground around the bushes. The key to control is an early spring cultivation that covers or destroys the fungus on the ground. As with most fungal diseases, overfertilization will encourage mummy berry. Organic gardeners seldom need to worry on this point.

Avoid high nitrogen fertilizers if botrytis or gray mold blight hits blueberries in your area. And prune regularly. The sign of botrytis is when young berries shrivel and turn purplish in color. Shoot tips die back, sometimes showing a gray mold on affected parts.

Because it is quite common, powdery mildew is usually considered a serious blueberry disease. However, the whitish mold on upper surfaces of leaves appears only after harvest is over. How much it affects yield and vigor is still in doubt —which means the disease has little effect. Berkeley, Earliblue and Ivanhoe seem most resistant among the newer varieties.

In the South, stem canker used to be a very serious disease until more or less resistant varieties were developed. The canker looks like, well, the stem looks like it has a bad case of poison ivy. Eventually, the plant dies, but it may take several years. New rabbiteye varieties are immune, so with them stem canker is no problem. Of highbush varieties, for the South only, Murphy, Wolcott, Angola, Croatan and Morrow possess resistance. And scientists are working on exotic crosses between highbush and rabbiteye varieties that may prove to be even more resistant, and have other advantages as well.

There are many other virus and fungal diseases, not to mention bacterial diseases, but if you plant good stock from reputable nurseries, keep your bushes pruned and your soil fertile and weedless, I daresay you won't learn about them the hard way. If you do have trouble, you'll have to consult an expert anyway.

Rabbiteye, Lowbush and Other Varieties

So far, I've talked mostly about highbush blueberries, the ones I know a little about. The other type that is cultivated for commercial and garden production is the rabbiteye variety—named as you can guess, because the berry does look, to some people anyway, like a rabbit's eye.

Rabbiteye varieties, like highbush, came originally from wild blueberries—ones native to northern Florida, Georgia, and Alabama. The first cultivated rabbiteyes were bushes selected from the wild and transplanted to commercial fields. Systematic breeding work began only in 1940 and the development of satisfactory commercial varieties is still very much in its infancy. Some of the better known varieties are Coastal, Garden Blue, Callaway, Tifblue, Homebell, Menditoo and Woodward, with new varieties being introduced regularly.

With the introduction of these varieties, the rabbiteye production area has expanded in the South. Since they can stand heat and drouth much better than highbush and need much less chill period during dormancy, the rabbiteyes seem to do well in any acid soil across the South, even on the coastal plain of Texas.

Southern gardeners should by all means plant the rabbiteye rather than the highbush. Grow more than one kind, too, because rabbiteyes are even more self-sterile than highbush.

Horticulturists see a great future for the rabbiteye in the South. It will grow better on upland soils than on low coast lands. It will take sandy soils but doesn't mind clay soils. It prefers good drainage but will stand some poor drainage.

Rabbiteyes should be planted at least 6 feet apart in rows

The blueberry is a newcomer to the garden; the fruit existed only in the wild state until early in this century.

12 feet apart. The bushes grow much taller than the highbush (which makes one wonder about names). When setting out bushes, mix about a bushel of peat moss with the soil you tamp around the new bush, and it will love you always.

Care of rabbiteyes is about the same as the highbush, but little pruning of the former is necessary. Just keep the bush from growing too thickly to let light into the center.

Stem canker doesn't bother rabbiteyes, as I have said. Leaf rust is another blight that can injure highbush berries much more than the rabbiteyes. In fact, while the same insects and diseases are generally a threat to both types of blueberries, the rabbiteye in its southern habitat seems better able to fend off attacks naturally than the highbush.

Ripening time for rabbiteyes is May and June. Harvesting has been mostly hand-picking in the past, but mechanical harvesters will likely take over the bulk of the job as the rabbiteye becomes more of a commercial berry. Once ripe, rabbiteyes, like highbush berries, will hang on the bush a week without harm. You can't beat that.

Rabbiteyes can be propagated by softwood cuttings, as described earlier. Hardwood cuttings root less frequently.

Rabbiteyes will endure a wider range of soil pH and that fact alone makes researchers believe that a big expansion of the rabbiteye industry lies ahead. And it's all good news for southern organicists and berry lovers.

Of the many wild sub-species of blueberries, only 3 have commercial importance now, and of the 3, the lowbush blueberry is far and away the one most often marketed. The lowbush is harvested throughout the Northeast, but especially in Maine, and also in northern Michigan, Wisconsin and Minnesota, plus vast areas of Canada. It likes open rocky uplands and sandy "barren" areas where one would think nothing of value could be grown. The wild plants spread by underground root rhizomes, from which the new shoots grow.

The greater the number of shoots and the more vigorous they are, the better the crop will be. Severe pruning encourages good shoot growth—pruning is done traditionally by simply burning off the old bushes. The Indians first learned that trick: Burning the current growth to the ground brings a strong shoot growth the next year, then a good crop the year following. Burning also helps control weed and brush growth and no doubt kills some insect enemies too. So far neither mowing nor spraying herbicides does quite as good a job as burning.

Blueberry "fields" are either burned every 2 years or, in some cases, every 3 years, if the grower decides to harvest a second crop from the same bushes. The plants must be burned in late fall or very early in the spring before any new growth begins. If the ground is frozen, little harm is done to the plants. Straw or hay or some other highly inflammable material is spread over the field to make a fast clean burn, though machines with oil or gas burners pulled by tractors are now much more commonly used.

The lowbush blueberry suffers from most of the insects that bother highbrush varieties, but has a few all to itself. The black army cutworm is one of these. It feeds on buds, especially buds from new shoots after a burning, and leaves the buds browned as if from freezing. Another worm, called the W-marked cutworm (now there's a name for you) is very destructive in both Canada and Maine. The white-marked tussock moth has become a pest. Its first known outbreak of economic importance was in 1956, some years after the general introduction of chemical sprays like DDT, which makes one wonder if perhaps the sprays themselves caused the imbalance in the tussock moth population.

At any rate, organicists who live where the wild lowbush blueberry grows have a certain source of berries every summer with no more work than the picking. The wildling will usually come in by itself wherever forests are cut down or fields abandoned. Along the trail that leads to the lookout points on Hawk Mountain in eastern Pennsylvania (a favorite haunt of birdwatchers interested in the fall hawk migration), merely cutting down trees in one area of the forest allowed enough sunlight to come in so that wild blueberry bushes sprang up as if by magic.

If you think you have soil that is acid enough to grow lowbush blueberries and you want to get them started there, the best way is to dig up chunks of dirt about 6 inches in diameter and 6 inches deep in an established planting. The spadeful of dirt will contain segments of the rhizomes from the plants roundabout. Professionals call that a chunk of "blueberry sod." Transplant it and new shoots will come up, usually at the point where you cut into the rhizomes in your digging.

While mechanical pickers are beginning to get into the harvesting act on lowbush blueberries, professional hand pickers still do most of the work using a blueberry rake which is somewhat like the better-known cranberry rake. The picker scoops forward with his rake—nothing more than a set of long tines or fingers set close together—then lifts up and tilts the rake back. The blueberries are pulled free of the stems by the action of the tines. The berries must then be winnowed to blow away the stem pieces and leaves that come off in the rake.

Three other kinds of blueberries are picked from the wild and sold at market. One is the evergreen blueberry of the Pacific Northwest, which makes an attractive ornamental. Its branches are highly prized by florists for floral backgrounds in bouquets. More money is pocketed from the sale of branches than from the sale of berries. In the same general area of the Northwest, but farther inland, grows the mountain blueberry, which is usually picked casually, the way we gather wild strawberries in the East. Then there's the dryland blueberry, *Vaccinium pallidum* (just to impress you with how I can throw those Latin phrases around), which grows in the Southeast and up into West Virginia and southern Pennsylvania. Though wild, the harvested dryland crop has an annual estimated value of around $300,000, which is nothing to sneeze at. The evergreen and the mountain blueberries each average around $200,000. To get some perspective on that, the lowbush crop in the U.S. alone is valued at well over $5 million. The highbush crop ought to be above $20 million by now.

Organic farmers, especially those going back to the land, should investigate whether they can get a piece of that action.

Chapter Five

The Remarkable Blackberry Family

Over a hundred kinds of blackberries grow worldwide, and they're all good folks to know. The domesticated branches of the family are composed of 4 more or less distinct groups of kissing cousins: the upright bush blackberry, the trailing blackberry, the black-fruited dewberry and the red-fruited dewberry. The last group is the weirdest of all, as it contains kinfolks whose relationship to the rest of the family isn't too clear. The loganberry, the boysenberry, the youngberry and the nectarberry all may have some tomcatting red raspberries among their ancestors. The loganberry and boysenberry are named after the men who developed them, but it is not entirely clear with the loganberry whether Mr. Logan found the red blackberry or crossed a dewberry with a red raspberry. If you're like me, you don't really care. All these berries taste good. The important thing is, can you grow them?

You can grow at least one of the four kinds just about every place in the United States. But there's even better news than that—you might not need to. The blackberry is one example of a fruit that grows wild almost everywhere *and* with a quality often equal to domesticated varieties. If you live out in the country or have access to country fence rows and brushlands, you can usually harvest an adequate supply of blackberries without the work of growing them. My father-in-law picked 6 *gallons* of them in one day this summer (1973) from thickets on his farm. Back in the days when he was raising a big family and times were hard, those berries were picked and sold at market in Louisville, Ky. So heavy was the yield on those wild canes that the family could pick 4 to 6 crates a day during the height of the season—that's 24 quarts to the crate. Of course, at that time a crate sold for a disheartening $1.25. Today the blackberries still grow thick and there are people who will pay 50¢ a quart for good, ripe blackberries. That's $12.00 a crate. But as granddaddy says: "There aren't many folks willing to work that hard at the picking."

For one reason or another, many people don't have the opportunity to hunt wild blackberries or prefer the convenience of having this easily grown fruit near at hand. It is for them that most of this chapter is written. But bear in mind that all the praise I am going to lavish on blackberries is meant for the wild ones as well as the tame.

Blackberries have been not only dessert for men for centuries untold, but medicine also. A tea brewed from the roots and leaves will cure dysentery, according to the folklore of both Europe and America. The berry itself has in my experience a similar benefit—it makes a good dietary balance to the laxative qualities of sweet corn, tomatoes, peaches and other fruits and vegetables I am loading up on about the time blackberries get ripe.

Meeting the Family-Members

The blackberry resembles the black raspberry, at least from a distance. However, where the latter resembles a thimble when picked, in that it is hollow, the blackberry has a core to which the fruit drupelets remain attached when the berry is picked. The quality of a blackberry is determined by size and core—the bigger the berry and the smaller the core, the better the quality. When you chew up a really good blackberry, the core is all but indetectible, and the seeds are soft and unnoticed rather than gritty. On low quality berries, especially some wild ones, the core is hard and green to tooth and taste. Berry gatherers soon learn to pass up the harder berries for the juicy bigger ones.

Dewberries, both black- and red-fruited varieties, usually are larger and longer than bush or trailing blackberries, and often of better quality. Boysenberries, for instance, will reach a length of an inch and a half or more, and the quality is unexcelled among blackberries. However, I personally prefer the taste of a regular blackberry to that of a boysenberry.

Bush blackberry canes are much stouter than dewberry canes; the latter will seldom grow upright but must be trained to a trellis. Bush blackberries are generally hardy everywhere in the U.S.; dewberries are not and must be protected in winter if grown in the north.

The bush blackberry is most widely grown in this country because it will take almost everything old man winter throws at it. Two of the newer varieties are Ranger, developed in Maryland, and Darrow. Darrow is a good northern variety; commercial growers look so favorably on it that the New York State Fruit Testing Cooperative Association offers no other variety for sale and has dropped propagation of Bailey and Hedrick. Ranger is especially suited for wine-making.

Eldorado is another variety favored by northern growers and is probably most commonly grown in Ohio and surrounding states. Early Harvest is a kind you'll find in the cornbelt, along with the Snyder variety. In the South look for Humble and Lawton. Alfred, being especially hardy, is recommended for the coldest northern states. Raven is grown mostly in the middle states.

These ripening blackberries will be a good source of vitamin A; a small serving (¾ cups) contains roughly 300 units.

Because bush blackerries are well-armed with thorns, gardeners and commercial growers alike appreciate varieties like Thornless and Smoothstem from which plant breeders have eliminated the ouch. Neither Thornless nor Smoothstem are extremely hardy, and you may have considerable die-back in zone 4 some years. They may kill altogether in zone 3 or colder.

Trailing blackberries are grown domestically principally in the milder parts of the Northwest—the Evergreen and the Himalaya being the old traditional varieties. Cascade, Brainerd, Chehalem, Marion and Olallie are other old standbys. These varieties are not as winter hardy as bush blackberries. Many crosses between bush and trailing blackberries have been made over the years, to further complicate the picture. Among the "semi-trailing" types are Jerseyblack, grown along the East Coast principally, and McDonald, which you'll likely find in Texas.

Black dewberries are one of the few commercially grown crops that actually seem to prefer a poorer soil. Well, I'm not sure "prefer" is the right word. But they will produce a crop on soil too poor for other berries. Organic farmers with a short supply of natural fertilizers, please take note—if you live in the South. Black dewberries aren't winter hardy enough to be grown commercially in the North. Actually, few dewberry producers remain, even in the South, but the two I know about in North Carolina have all the business they can handle. Lucretia and Mayes are two old favorite varieties.

The most famous of the red dewberries today is the boysenberry, marketed so astutely by California growers. Some gardeners try to grow them in the North, but that's not very practical. I have a few myself and believe you have to like boysenberries very much to go to the trouble of protecting them over winter. More on that later.

Establishing Blackberries in the Garden

For me, choosing a blackberry variety for my garden came naturally—and I mean that literally. In this climate, my choice had to be a hardy, upright bush variety, and the one I settled on was growing wild in a nearby wood. I had found one small patch of berries of a very good quality compared to other wild ones in the vicinity. Since they were doing so well without any help from man and despite much competition from other brushy growth, I figured that grown lovingly in the garden, the berries would even be larger and juicier.

In early spring, I dug 4 of the canes up (having marked them the previous summer by tying a strand of twine around each so I'd be sure to get the right ones when I returned in March), carefully placing them in a cardboard box with as much dirt remaining around the roots as I could handle. I set them gently about 4 feet apart in a garden row I had tilled thoroughly the autumn before. And that's all there was to it.

The first year, the transplants grew accustomed to their new surroundings, sending up only 1 weak new shoot each. I cut the old cane off at ground level as soon as the new one got growing well, though I am not too sure that was a smart thing to do. My figuring was that the strength of the plant should go into the new cane rather than being divided between the old and the new, but that rationale may be just a fanciful bit of anthropomorphism on my part. Perhaps with both old and new canes growing, the root system would develop faster. I don't know. At any rate, the bushes grew only slowly until the third year when they suddenly took hold and began to branch and sucker vigorously. Now they are growing so fast that they will quickly take over the whole homestead if I let them.

Bush blackberries grow quite like red raspberries in that they multiply by suckering normally. (Some kinds will also tip root the way black raspberries do.) Mine do not send up very many suckers, not nearly as many as the red raspberries do, for which I am eternally grateful.

I allow suckers to come up in the row until the canes are spaced about 1 per foot—not quite as thick as I keep the raspberries. In the beginning, when suckers did not come in the row as fast as I wished, I would sometimes bend a low cane over and bury a portion of it about 6 inches back from the tip. A Y-shaped stick thrust upside down over the cane held the portion under the dirt, at which point roots would develop, starting a new plant. Then I'd cut it free from the mother plant.

When commercial growers want to produce a large number of blackberry plants in a hurry, either for berry production or for selling to other growers, they may start plants from root sections. Blackberry roots, from a pencil diameter on up, are cut into 2-inch lengths. The pieces, planted in flats or in the nursery (the former is preferable), will develop plants if placed an inch or 2 deep in moist soil. These plants can later be set out in the row.

Care and Feeding

My blackberries (and most upright bush types sold by nurseries) have stout canes. When they are about 5 feet tall in midsummer, I tip-prune them. As is somewhat true of black raspberries, the canes will then branch vigorously. As is not somewhat true with black raspberries, the blackberry canes will become stout and strong enough to stand through the winter without benefit of staking or trellising. In early spring, after freezing weather is past, I will prune back the side branches to within about 10 inches or less of the main cane. That makes a nice bushy plant from which you can pick the berries without being engulfed by those vicious and rapacious thorns. I ruthlessly rogue out suckers not in the row, so that row width stays about 10 inches wide, easy to pick along, cultivate by or mulch around.

I have kept the ground under the bushes mulched the whole year round and have not yet had to cultivate. Blackberry roots grow very close to the soil surface, so if you must cultivate them, do so only shallowly. A disk or rotary tiller is safer than a shovel cultivator. Old timers used to cultivate their blackberry fields until the middle of August. They thought cultivation beyond that time would encourage fall

Each of these dewberries is a mouthful in itself. To make harvesting easier, they should be trellised.

growth when the plants should be hardening off to resist winterkill. I personally don't believe cultivation will encourage late growth as much as mulch and wet fall weather. Besides, if you have tended to your weeds through the summer, you'll have little need to cultivate beyond the middle of August anyway—at least in the North.

Blackberries fruit on the canes that grew the previous year, as do raspberries. Do not pick the berries for table use until they are dead ripe. If you intend to make jam or pie you don't have to be so particular in this regard. Dead ripe, a blackberry is sweet and just as good as a raspberry. Sometimes better. Being larger than raspberries, blackberries pick faster too. An acre of them can easily yield 6,000 quarts.

Blackberry canes are sometimes trained to wires as I described for raspberries. But if the canes are tipped back to 4½ or 5 feet, they will not need support, in my experience.

Dewberries and trailing blackberries are a different story. Their canes ramble all over the ground, reaching lengths of 15 feet or more in a summer's growth. They increase by tip rooting, like black raspberries, rather than by suckering.

First year canes may be allowed to crawl along the ground. The second, fruiting year, the vines are lifted and tied to stakes or trellises.

Do the same with red dewberries—boysenberries and loganberries. Dewberries are not hardy in my climate, so I cover my low rambling canes with a foot or so of leaves over winter. In spring I tie them to a wire about 3 feet off the ground.

Immediately after the crop is harvested (dewberries ripen conveniently between raspberry and blackberry crops), I cut out the old canes, just as I do blackberries and raspberries. Such scrupulous sanitation is the best defense against disease and some insects.

In the South, where dewberries are more at home, common

practice in the past called for hoeing the old vines off just below the surface of the soil as soon as the crop had been harvested. With the longer growing season of the south, the new canes would then grow up to 2 feet tall before winter and did not require staking. The old plants were burned immediately after they were hoed off to control anthracnose.

To protect dewberries from winterkill in the north, some gardeners go to much more elaborate pains than I do. The most popular form of protection is to sandwich a bundle of canes in corn stalks, tie them securely, then lay the whole on the ground and cover with leaves or some other mulch. Obviously, dewberry patches will remain small in the North since the labor involved for winter protection is quite time-consuming if not arduous. I definitely would do it only for boysenberries; black dewberries are just not that much better than the more easily grown upright blackberries.

Insects and Diseases

Insects that attack blackberries and dewberries are those already discussed as enemies of raspberries. Chief pests are the red-necked cane borer and the raspberry crown borer. The former makes those cigar-shaped swellings on the canes usually near the base. Prune off the damaged canes and burn them when you do your early spring pruning. The crown borer shows no clear evidence of its presence: cane growth simply is weak. But when you cut old canes out after harvest, if they are hollow at ground level, likely as not crown borers did it. The insect is difficult to control in trailing blackberries, which it seems to prefer. Fortunately, it seldom reaches harmful population levels in upright blackberries or dewberries. The adult looks something like a yellow jacket with semi-transparent wings. The eggs, laid on the underside of leaves, resemble mustard seeds. They hatch out in the fall.

Where I live, orange rust is a common disease of blackberries. At least I see it quite often on wild plants. The striking, bright-orange blight spores appear during summer on the undersides of the leaves. New shoots, not yet so affected, will nevertheless be quite spindly and unproductive. The only control is to grub out infected plants—roots and all—and burn them.

Anthracnose can be bad on black dewberries, I understand, but since I don't grow them, I don't know how much of a threat the disease is. Since not many large commercial fields remain today, I doubt anthracnose is a big worry. Control it organically by cutting out the bearing canes as soon as possible after harvest.

Double-blossom is another blackberry disease I've become acquainted with only through reading. To control the fungus, break off reddish, spongy buds that produce the double flowers and burn them.

This crate holds two varieties of loganberry, ready for the market: the logan and the mammoth blackberry.

Virus diseases can attack blackberries just as they do raspberries, and wild blackberries are often infected. There's no control for viruses, but you will go a long way toward avoiding them by buying good stock from good nurseries. Some nurseries (Bountiful Ridge of Princess Anne, Md. for one) offer registered virus-free blackberries, as they do with raspberries.

Further safeguarding requires that you grub out infected wild plants in the vicinity of your garden or field. At least that is what the experts advise. I have to wonder about that a little, since my plants came originally from the wild and were growing good berry crops though other wild plants nearby were obviously infected with some disease. I figure (hope) the plants I put in my garden have some special resistance. So far, so good.

After the Harvest

There are recipes in most good cookbooks and specialized fruit and dessert cookbooks to help you turn blackberries and dewberries into delicious jams, pies, cobblers and wines—besides eating fresh with honey, cream and shortcake. However, you may not have heard of blackberry jam cake—which would be something of a minor tragedy. My wife makes it following the directions her mother uses to make it, following no doubt a recipe her mother used. Each generation has added another touch to it, and I doubt whether the recipe appears in any cookbook exactly as they concoct it. But everyone ought to have a piece of "jam-cake" once in their life anyway. Here's how to make it:

1 cup shortening
2 cups raw sugar
5 egg yolks beaten
4 cups sifted flour
1 teaspoon cloves

1 teaspoon nutmeg
1 teaspoon cinnamon
4 teaspoons baking powder
½ teaspoon salt
1 cup coffee (ready to drink)
½ cup wine
1½ cup of blackberry jam
1 cup raisins
1 cup black walnuts
5 stiff-beaten egg whites

Thoroughly cream shortening and sugar. Beat in egg yolks. Sift together dry ingredients and add alternately with liquids, beating well after each addition. Mix in jam, raisins and nuts. Fold in egg whites. Bake in three paper-lined 9-by-1½-inch round pans in moderate oven (350 degrees) for 30 to 35 minutes.

Frost with caramel frosting. Soak a rag with bourbon, wrap it around the cake and store in an airtight container until ready to use. A jam cake can be stored that way indefinitely —as long as any fruitcake can be stored.

I have to make a comment about this recipe since I've never read the observation in any other cookbook. If you go back to the recipe above, you will notice that the first word is "thoroughly." Now when my wife or my mother-in-law says "thoroughly" she means that most emphatically. I wish you could see either one of them during that first beating process on the way to making a cake. "It's the only way to make a really good cake; the secret of it," claims Mother-in-Law. For a good jam cake, the shortening (butter) and sugar must be creamed by hand-beating until the mixture is as light as thistledown and as smooth as gelatine.

Good blackberry hunting—and don't forget, if you live where chiggers are bad, smear bruised mint or pennyroyal leaves on your neck, wrists and ankles before going to the woods to pick blackberries.

Chapter Six

Mulberries and Elderberries

You may wonder why I mention 2 such different berries in the same breath. Mulberries grow on trees and ripen in early summer; elderberries grow on bushes and are ready to eat in late summer.

The 2 berries do have something in common though. They *both* can provide you with good desserts, though many people don't know that, and, more importantly, both afford excellent food for birds. In the latter case, the 2 berries often perform a significant service for the wise fruit grower by luring birds away from his more treasured berries. I have observed too often to have any doubt—in years when mulberries are plentiful, the birds eat far fewer blueberries. When elderberries are available, birds don't bother blackberries and late raspberries as much.

The Mulberry Tree, and Why Great-Grandpaw Planted It

Humans don't like mulberries as well as birds do. Most mulberries have a bland, insipid taste, which no doubt explains why there's presently very little interest among gardeners in growing the fruit. Secondly, larger mulberry trees (and they can grow as large as a maple) produce a prodigious amount of fruit, half of which often ends up as an unsightly mess on the ground.

There are solutions to both these "failings" (which I will get to later), but even if there were not, I would still value the way mulberries protect my other fruit from the birds. I strongly recommend that the gardener who loves berries, but loves birds also, get a mulberry tree started as soon as he can.

There's a big tree right at the end of our driveway, fortunately for me. I suppose it has been there at least 50 years, guessing from the size of it. A dozen more mulberries grow along the road between here and Driscoll's barn. These trees are of varying age, one quite young, sprung up, I presume, from a seed carried by a bird. I'm quite sure no one planted it. The older trees were probably started by turn-of-the-century gardeners who knew something about the ways of birds.

About half our trees produce black mulberries and half produce white ones. The blacks taste a little better than the whites. Neither, however, tastes like the mulberries from the small tree—hardly more than a bush—that grew on the playground of my grade school in Ohio. Maybe I was just hungrier then, but those mulberries seemed sweeter by far than our present mulberries.

It is difficult to find much information about mulberries (beyond their role in feeding silkworms), but what I have learned lends credence to my memory of those mulberries of my childhood. According to the berry volume of the "Biggle Books" (a rare set of garden and farm books published intermitently around the turn of the century by the Wilmer Atkinson Co., Philadelphia) mulberries even in 1900 were "nowhere grown for market purposes." However, the book continues: "The Downing mulberry has real merit, but is not quite hardy in very severe climates."

Remembering that item, I was delightfully surprised to

These curious-looking white mulberries don't please everybody's palate—the fruit is at its best in jams, and in pies and other desserts.

learn that the New York State Fruit Testing Cooperative Association at Geneva, N.Y., is now offering for sale a mulberry it calls the Wellington, propagated from an old tree near the experiment station. The catalog says the Wellington may be "an old variety, New American, which was also sold many years ago as Downing."

The catalog adds: "Mulberries from commercial sources are usually inferior types of little value for fruit production. Wellington's fruits are long, slender, cylindrical, soft and of good flavor. This is the best mulberry known at the Experiment Station. The tree produces heavy crops and the fruit ripens over a period of several weeks."

Two deductions seem to be in order. Whether Wellington derives from the old Downing or not, it has to be fairly hardy to survive New York winters. But the fact that the Wellington was found so handily next to the experiment station could hardly be coincidental. Plant breeders just haven't looked very hard for high-quality mulberry trees; there are probably many more with equally good fruit scattered around the country.

Mulberry trees you spot near old barns or homesteads most

likely mark the place where the hog lot or the chicken run was once located. Why? Both chickens and hogs like mulberries, and practical homesteaders of the past planted the trees where they would get the most use out of them—food for domestic animals, wild birds and themselves—a hint to modern homesteaders. If you set out mulberry trees to make food producing shade for hogs, space them at least 30 feet apart both ways. (Better to put a hog lot among acorn producing oak trees and interplant the mulberries.)

Chickens with access to mulberries can be allowed to roam your garden for insects. The fowl will gorge on the fruit, then not bother fruit or tomatoes much in the garden. (If only we had preserved *all* the wisdom of our ancestors!)

Ancient mulberry varieties include the white mulberry or Chinese mulberry, used principally (the leaves) as food for silkworms. There is a so-called Russian mulberry, its origin lost in antiquity, with red fruit and very hardy. Derived from the Russian mulberry is the weeping mulberry which used to be a popular ornamental. Then, in the eastern half of the country, there's a native American or red mulberry, which bears a purplish red fruit. From this type, plant breeders in years past developed the more flavorful cultivated varieties: Black English, Downing, Hicks and New American.

Mulberries will grow just about anywhere, as they tolerate a wide variety of soils. Neither insects nor diseases bother them critically, probably because there is no commercial concentration of them to support a bug or blight buildup. There is a canker disease that affects the trees, controlled by cutting out affected branches a foot below canker signs during winter. In the South, a curious ailment known as popcorn disease sometimes strikes trees, but its effects are not serious. Affected fruit will not ripen properly, but the disease seems to go away of its own accord. Mildews may occasionally kill leaves, but will not seriously harm the tree.

The easiest and fastest way to pick mulberries is to lay a

clean sheet under the tree and shake the limbs. Ripe berries fall easily. The berries can be frozen as well as used fresh, for pies.

Mulberry lovers swear by a pie filled with a mixture of ⅔ mulberries and ⅓ rhubarb. The latter adds a tartness which mulberries lack. Gooseberries mix well with mulberries too.

The so-called French mulberry is an ornamental bush, not very hardy, and is neither French nor a mulberry. The paper mulberry is a sort of second or third cousin to the true mulberries. Its fruit is not eaten. You see it growing sometimes along city streets, since it seems impervious to city pollution and stress. But it will spread and become a nuisance in gardens. The tree gets its name from its papery bark.

The Elderberry, and Why Grandpaw Didn't Plant It

The elderberry holds a treasury of faraway memories for me besides a treasury of present day good tastes no farther distant than the hedgerow outside the window where I work. First of all, I'm reminded of Aunt Stella's elderberry pie—the legendary one she baked every year for her boys Tom, Edgar and Charlie, which according to *them* was "half as big as a washtub." Loving elderberry pie, I hoped wistfully each August to be invited to such a feast. Alas, I never was.

Then there was the elderberry jelly which Maggie, who lived down the road from us (Aunt Stella lived a mile across the fields), made with other kinds of fruit mixed in. She called it elderberry delight, and so it was.

The pure, undiluted kind of elderberry jelly reminds me of my sister Berny, not only because she makes it every year, but because when she was born, the woman who came to keep house for us while mother was abed brought with her the knack of making the stuff. It was so good, mother made it from then on.

But most of all, elderberries remind me of my grandfather,

Henry Rall, who was what I could only describe as a farmer's farmer. To Grandpaw, any other plants besides corn, wheat, oats, hay and strawberries were weeds. Grandpaw hated weeds. He fought them in hand to hand conflict all his life and even after he turned over his farms to his sons and sons-in-law, he continued to fight them. He'd come almost daily from his house in town and walk the fields with a spade, his weed killer. With it, he could deftly fell any weed, no matter how big, with one expert thrust.

But there was 1 weed he never got the better of—a big elderberry bush along the road in the fence row near where the home of 1 of my sisters stands today. Grandpaw thought a farmer lazy who allowed weeds to grow in a fence row. That included elderberries, despite protests from us. "Let them things go and they'll take over the whole field," he'd huff and wade into action, spade lashing at the stout and stubborn elderberry shoots.

He may have been right too, for he never did kill the bush. He'd whack it off every summer but that only seemed to encourage it to grow back more vigorously. In later years, Grandpaw slowed down a bit and didn't get his 40 whacks in, which allowed us and the birds to harvest the berries. Until weedkilling sprays came along, that is.

Cultivating Elderberries

Two distinct types of elderberries are native to the United States, one in the East and Midwest, the other in the far West. The latter is blue and seldom eaten as far as I know; the former is almost black when ripe and delicious in pies and jellies to most folks' taste. Few plant fanciers have turned to the elderberry for ornamental use, probably because of its rank growing habits. The bush's creamy white blossoms are beautiful, however, and the cut-leaf varieties are quite handsome.

The eastern elderberry (the one I know about) grows very vigorously if moisture is ample. It grows even more vigorously in the Midwest than it does in the East in my observation. New shoots grow out from the base of the plant each spring and may grow 7 feet or more the first year. The next year those shoots fruit and will continue to bear for 3 or 4 years, after which they weaken and eventually die out as more shoots grow up around them. The bushes can spread from shoots or from seed. And spread they will. You won't have to encourage them.

Since each clump of plant will grow to several feet in diameter and 12 or more feet tall, I wouldn't plant them in a cultivated garden. Rather I'd establish them in fence corners, or along a lawn's edge. Full sunlight is best for them, though a little shade won't hurt. The plants like to be well watered but do not grow in poorly drained areas. I think the best place for them is in hedgerows along with other berry plants birds like. (More on that in Chapter 10.)

Tip back new shoots at the end of their first growing year about a foot. This will make the shoot put out side branches and fruit better. Cut out 3-year-old wood. Chop off new shoots growing too far out from the bush, unless you want the plant to spread.

If you want to make 2 or more elderberry bushes bloom where only one had bloomed before, you can dig up the plant and divide the clump of roots into 2 parts (or more) and replant. Make sure the transplants get watered well during the first year of growth.

I didn't have to plant my present bushes at all. They just appeared one summer, 1 in the hedgerow on the north side of the property and 1 at the wood's edge on the east side. No doubt birds carried the seed from other wild plants. All I do now is trim back the wild rose, honeysuckle and bittersweet that threaten to engulf the elderberries and the bushes yield well every year.

Commercial elderberry growers (there are some, believe it or not) follow a regular schedule of fertilizing. But in small scattered plantings, I have never met an elderberry bush in my life that needed fertilizer to make at least a satisfactory crop. I suppose there are bugs and blights that bother elderberries, but I've never noticed. The elder seems to be the berry most free of pests. That quality, plus its storehouse of vitamin C, makes the elderberry one of the organicist's regulars.

These ripe elderberries are ready to be raked off by a harvester's hand. Very high in vitamin C, elderberries are used to make jam, pies and wine.

Nurseries maintain that elderberries are to some extent self-unfruitful. Since I've seen plenty of bushes fruit well all alone along corn field fence rows, I'd like to argue that point, but maybe there's something to it. As long as you're getting 1 plant, you might as well get 2.

Adams is one of the better known varieties. It's a selection from the wild made about 1920 in New York. Another older variety is Ezyoff, referring to the berry's propensity for coming free from the stem readily when ripe. Ezyoff has been crossed with Adams to produce "New York 21," offered by the New York State Fruit Testing Cooperative Association at Geneva. (I realize I mention this particular nursery frequently. I have no money in this enterprise nor do I wish to imply that plants from this nursery are any better than any other plants. I just happen to buy from this source. Because the nursery is closely connected with the state university and the state agriculture research stations it has always impressed me as a no-nonsense source of dependable stock. Sometimes, however, supplies are limited.) York and Nova are two more varieties, generally considered better than Adams for most areas.

In harvesting the fruit, pick *whole clusters* of the small, shot-sized berries into a basket. Later, in the cool of the shade, you can strip the berries from the stems. Gently. You should try very hard not to get any (or many) small stem pieces in with the berries, as too many bits of stem can give your pie, jelly or wine an off flavor. If the berries are dead ripe (black in color) they will strip off fairly easy. Trouble is, if you wait too long for the berries to get fully ripe, the birds will win the whole ball game. They are adept at picking out the ripest berries in the cluster too.

My elderberry wine never turns out very well, though I have tried to make the stuff many times. Mine is always a little reminiscent of coal oil and moth balls. But I've tasted the wine made by people who know the mysteries of such things, and it was delicious.

Wine can also be made from the elderberry blossoms, just as a wine can be made from dandelion blossoms. Winemakers call them elder-blow. Also, elderberry blossoms dipped in batter and fried a golden brown are supposed to be a real gourmet's delight. Every year my wife and I vow to try this dish, but never do. I think it's because elderberries blossom at a very busy time for us in the garden.

Elderberry sauce can be used in much the same way as cranberry sauce (and is also made the same way). Maggie, our neighbor of long ago, used to say that everyone should boil down elderberry juice to a syrup to take in winter to ward off colds and soothe coughs. Since the berry does contain considerable amounts of vitamin C, I suppose you can't go wrong.

Elderberries are quite seedy, which bothers some folks. But if you fill a pie with half apple and half elderberry, the seediness is not so noticeable, and the distinctive elderberry flavor is still retained.

If you want to hunt wild elderberries, try looking along back country roads in mid-August. You should find them anywhere brush is growing in the open (not under large trees) along creeks or fencerows. There are other berries somewhat similar to elderberries which are decidedly inedible if not poisonous. If you aren't *sure*, show your berries to someone who is, before you eat them.

Chapter Seven

Currants and Gooseberries: Cold Climate Friends

I used to react to the idea of growing gooseberries and currants with the kind of righteous alarm that comes so often with only a little knowledge. "What? Grow *Ribes?*" (The use of the proper scientific family name of the two closely related berries was calculated to chase away the last lingering doubt in the minds of my listeners.) "You can't do that. Those two plants are hosts of a disease that could wipe out the white pine trees of America!"

Not altogether surprising to those who know me well, I was exaggerating, and not exactly slightly either. The disease, white pine blister rust, spends part of its life cycle on certain varieties of *Ribes* and the disease *is* dangerous to white pines. But the threat today is not nearly as serious as it is sometimes thought to be. Government regulations and years of research have seen to that.

You cannot, for instance, ship *black* currants across state borders, though why anyone would want to is beyond me. The reds and yellows make better jelly and they can be shipped into most states. In certain areas where white pines are an important commodity—where the trees grow in large forested areas—the growing of currants and gooseberries is prohibited and plant protection agencies have tried to destroy wild bushes. These areas, properly called White Pine Blister Control Areas, are all mapped out, and in states where they exist you may need a permit to grow gooseberries and currants—if you live near or in a control area.

Reputable nurseries relieve you of the worry about who can and who can't grow *Ribes*. If you live where you need a permit, the nursery you buy from will either have that permit or send you a form that you fill in and return for a permit. But more importantly, the types of red and white currants and gooseberries they sell will be free of the disease. (Viking, a red currant variety developed already in 1935, was considered completely immune.) As Malcolm C. Shurtleff says in his book, *How To Control Plant Diseases* (Iowa State University Press, 1962): "Plant only currants and gooseberries from a reputable nursery. These will not become infected by the blister rust fungus."

If you have white pines and want to be super-safe, plant your gooseberry and currant bushes at least 900 feet from your white pines. The fungus spores can't travel any farther than that alive. If you live near 1 of the control areas, state law may prohibit you from planting gooseberries and currants any closer to the area than 900 feet. Nor can you plant them within 1,500 feet of any white pine nursery. Regulations vary somewhat from state to state, so you would be well advised to check with your nearest USDA office or with your local Extension Service office. Or you can write USDA, Agriculture Research Service, Plant Pest Control Division, Washington, D.C. 20251. But for most states, the prohibition against *Ribes* (other than black currants) does not apply. The states without the prohibition are: Alabama, Alaska, Arizona, Arkansas, California, Colorado, Connecticut, Florida, Idaho, Illinois, Indiana, Iowa, Kansas, Kentucky, Louisiana, Michigan, Mississippi, Missouri, Nebraska, Nevada, New Mexico, New York, North Dakota, Oklahoma, Pennsylvania, South Carolina, South Dakota, Tennessee, Texas, Utah and Wyoming. But it's important to remember that the other states are not totally sealed off from red and yellow currants and gooseberries; only control areas within these states are. However, without a special permit, you may have trouble getting a nursery to ship *Ribes* to you if you live in a state with White Pine Blister Control Areas. If you live close to extensive white pine forests, you will not want to take the risk anyway. No currant jelly, no matter how delicious, is worth risking a pine forest for.

Once past the white pine blister rust problem, who should be interested in growing currants and gooseberries? Everyone, jelly lovers who know currant jelly would say. Everyone, pie lovers who know gooseberry pie would say. Everyone, Englishmen who value gooseberries the way we favor raspberries would say.

But even pie lovers must face reality. Neither of these two fruits will grow in the deep South or on the hot plains of the West. They do like cold weather, however, or at least are extremely hardy to it. If you live where winter temperatures plunge well below zero, you can count on gooseberries and currants when all other berries freeze out.

Growing Gooseberries

There are two types of gooseberries: American and what is usually referred to as English. The latter varieties (there are a great many of them grown in England) do not do well in the United States, generally speaking. They are very susceptible to mildew. And they dislike our strong American sunlight. Fredonia, a red variety, and Chatauqua, a green, are somewhat of an exception. They can be grown fairly well (Fredonia better than Chatauqua) in certain areas: the eastern

Although popular in England, the gooseberry has yet to catch on in this country.

parts of zone 2 and 3 and along the Pacific coast in zone 4.

American varieties, on the other hand, are quite easy to grow. Poorman is a good red one usually offered by nurseries that sell gooseberries. Welcome and Pixwell are other favorites that ripen to a pinkish color. Green Mountain is a fairly reliable green or white variety. Downing, Oregon Champion, Como, Red Jacket and Carrie are among the oldtimers you might hear about.

The redder the variety, the sweeter the berry, and so the better to eat fresh. The greens make good tart pies and jams, for which purpose the berries do not necessarily have to be dead ripe. You can let ripe gooseberries hang a week or two on the bush without hurting them.

Picking gooseberries is not exactly what I would call a fun thing. Nasty thorns, with absolutely no regard for anyone, get in the way. Experienced pickers wear gloves, hold a branch in one hand and strip the berries off with the other. You get a few leaves in your basket, but you can winnow them out later.

A picked gooseberry has a stem remaining on one end and

a blossom collar on the other. Both should be pinched off when the berries are prepared for eating. A lot of trouble? Here's a gooseberry pie lover talking: "When I find gooseberries in a market, I buy every one in sight. I've got one quart left in the freezer which I'm saving for a special occasion. In fact I've been saving them for a long time because gooseberry pie is so good I'm having a hard time finding an occasion special enough to fix one!"

Gooseberries make nice garden bushes because they won't spread all over creation like blackberries or raspberries, nor do they grow rank and tall like elderberries. Sometimes, however, a few too many canes will come up, and you will have to thin them out for good air circulation.

Unlike bramble canes, gooseberry canes will produce fruit for 3 or 4 years before they begin to lose vigor. Usually it's best to cut them out after the third bearing year. The sign of a still vigorous cane is when new growth exceeds 6 inches a year. In general try to maintain about 9 canes per bush: 3 of them 1 year old, 3 of them 2 years old and 3 of them 3 years old.

Young, year-old canes may be tipped back to induce branching, but on a vigorous bush, branching may not be desirable. Remember that every gooseberry bud will produce 1 or 2 berries, possibly 3, but no more. (A fruit bud on a currant bush produces a whole cluster of berries.)

Get your pruning on gooseberries finished very early in spring. The plants break dormancy unusually early—almost as soon as the thaw is out of the ground. For this reason, some authorities advise fall planting. In the colder parts of the North, however, I'd wait till spring to plant my gooseberries.

If you buy an old country homestead in the North, you might discover some gooseberry bushes growing on it, untended for years. Most or all of the canes will be old, perhaps not bearing any fruit at all. Cut out all the canes except 2 or 3 of the healthiest. In this way, you will induce new shoots to grow, after which you can cut out the remaining old canes. In 2 years, you have a "new" bearing bush.

Gooseberries will grow in heavy soils better than any other berry will (except currants). They'll grow in sandy soils too. Nor are they particular about soil pH as long as the soil is neither extremely acid nor alkaline. Like all berries, the plants do best when mulched. Several forkfuls of manure around the base of a plant every summer is all the fertilizer it needs.

The easiest way to increase a gooseberry bush is to bend a low branch over and bury a length of it in the ground, allowing the tip of the branch to stick out. The buried part will put down roots, and a year later you can cut the branch from the bush below the new roots and transplant it. Sometimes, a low branch will bend over to the ground and root of its own accord. If you don't want more bushes, be sure to prune such laterals out, so the bush doesn't get too thick.

There are wild gooseberries too. Near where I once lived along the Minnesota River, I found them in an open woodland. Temperatures used to fall as low as 25 below zero in that vicinity—so cold our skates would no longer glide easily over the ice, but would squeak along as if we were trying to skate on tin. That's how I know gooseberries survive well in the North. Those wild ones were green, eventually turning whitish-yellow, but never sweet. We used to eat them anyway. But I prefer the domesticated red ones now.

Cultivating Currants

Almost everything I've said about gooseberry culture applies equally to currants only more so. The old currant bushes at the Ohio home of my childhood produced berries every year without any care whatsoever. But the one on the north side of the house, which received more shade, did better than the one on the south. Commercial growers used to understand that proclivity for shade and took advantage of it. They'd plant their gooseberries and currants between alternating rows of grapes or between rows of orchard trees to protect the berries from the unremitting sun of a long summer. However, I noted beautiful Red Lake currants growing on a Kentucky farm a couple of years ago without benefit of any shade at all. So one has to wonder. But at least experts all agree that in the *far* North, shading is *not* needed for a successful crop. On the other hand, both currants and gooseberries will stand some shading without ill effect, and that's good to know. There are not many food plants that you can say that about.

The leaves of a currant bush look about like those of a gooseberry, but the berry itself is quite different. The currant is roundish; the gooseberry oblong with telltale lines, running parallel to each other the length of the berry—like longitude lines on a map.

Currants come in 3 basic colors: red, yellow (sometimes called white) and black. No matter what older folks with roots in Europe say, I wouldn't waste time on the black currant. Not only is it in all likelihood the main culprit in spreading white pine blister rust, but it, well, it has a rather unpleasant odor to my nose. I think if you cook anything long enough and sweeten it, it might taste passable. Even an old pair of shoes. The same with the black currant.

Among the red varieties, Red Lake and Minnesota 71 are about the best, and they are usually available. Other suitable older varieties include Wilder and Perfection. Going way back, you'd run into Victoria, Fay's Prolific, Cherry and Red Cross.

White Grape is the old reliable green-white-yellow variety. A new one, Imperial, is now available from the New York State Fruit Testing Cooperative. Plant breeders at the Experiment Station there consider it "considerably better than White Grape."

The rules for pruning currants are the same as for gooseberries, though you do not always need to maintain as many fruit buds. Remember that each bud on a currant bush produces a whole cluster of berries.

When you want more currant bushes, take 12-inch cuttings in early spring from an established bush before it breaks dormancy. Stick each cutting about 8 inches into the ground where you want it to grow permanently. Water it well the first year. It helps to cover each cutting with a glass jar. This method will give you about a 50% survival rate—in other words about half your cuttings will root and grow.

For jams and jellies, currants are usually picked before they are dead ripe. I think gooseberries are better in every way if

The currant is a first cousin of the gooseberry. It is the prime ingredient in currant jelly, but it is not the "dried currant" available in supermarkets.

ripe, or nearly so, before you pick them. At any rate, don't pick either kind of berry until you are ready to use it. Extra ones at the end of the season can be frozen for use later. (The store-bought dried "currants" that look and taste like small raisins are *not* currants. They're raisins from a special type of grape.)

I'm not too sure if there is enough interest in currants to build up a market for them. But there seem to be quite a few people knowledgeable about gooseberry pie, and for sideline income you just might be able to work up a good trade. The highest yield I've ever heard of for gooseberries was 7,500 quarts per acre. That's a lot of pie.

Currants used to be mixed with red raspberries to make jelly. The currants supplied the pectin. I've not tried that myself, but it sounds good.

Pests and Diseases

Gooseberries and currants share most of the same pests and diseases. Powdery mildew is probably the most serious; it certainly is on European-type gooseberries. There's not a whole lot you can do to stop mildew, especially organically. Keep the bushes thinned of rank growth for good air circulation and don't plant the bushes where they will be *too* shaded, if mildew is a problem. Since the bushes like some shade where summers are long and hot, mildew can put you in a dilemma. If it does, stick to American varieties.

Anthracnose can be harmful to both currants and gooseberries. There's no cure, but some varieties are more resistant than others. "Welcome" is the most resistant gooseberry of all to this blight.

Various other cankers and rots may at times raise their ugly heads among your currants and gooseberries. Often, such diseases can be avoided simply by not fertilizing too heavily with nitrogen.

Insects are rarely a critical problem on currants and gooseberries in small plantings, but they can be pesky sometimes. If you look at your bushes one day and think you have discovered a weird new disease that gives the leaves a yellowish-reddish, warty appearance, you'll probably find knots of aphids under such leaves. If a currant bush is suddenly denuded of its leaves, usually it means a gang of currant sawflies has eluded your watchful eye too long. Handpicking can control these characters easily on a few bushes. On larger plantings, growers used to use an organic insecticide they called "hellebore" and which they considered toxic to bugs and not to humans.

Their hellebore must have been a very weak mixture and I would not advise anyone to try to make their own without some expert help. Hellebore is derived, if I am not mistaken, from a plant called false hellebore *(Adonis vernalis).* It looks much like skunk cabbage and grows in the same kind of low marshy area. The roots and leaves are *very* poisonous to humans if taken straight.

Borers may kill or weaken a cane or 2 periodically but are not usually serious on *Ribes.* Cut infected canes out and burn them.

Sometimes currants become infested with the small maggots of fruit flies. Organicists should dust with rotenone as soon as the blossoms have wilted, if they have had trouble with the maggot in previous years.

Mosaics and other virus diseases infect currants. Getting good stock from reputable nurseries is your best defense. And that is good advice, no matter what your problem, when and if you decide to add gooseberries and currants to your garden.

Chapter Eight

Grapes: THE Small Fruit

While the grape is not usually called a berry, by definition it deserves the name more than some other berries do. Horticulturists often head off arguments that might develop over this matter by referring to both berries and grapes as the "small fruits." That way only very correct scientists have to waste time deciding what is and what is not a berry. Technically even a banana is a berry.

Whatever, the grape is commercially the most important of all "small fruits" and perhaps all large ones too. The grape in its many varieties and uses accounts for a full one-quarter of all fruit production in the world.

Most of that production goes into the huge world-wide wine market, now growing in the U.S. by leaps and bounds. So popular are raisins, the dried fruit of special seedless grapes grown in the West, that marketers almost consider them a fruit in their own right. Grape juice enjoys a similar popularity and importance. And last but not least, dessert grapes for fresh eating continue to maintain high consumer demand.

Unending Varieties

There's a grape for every climate. In North America, Concord-type grapes, the slip-skin "eastern" grapes (*Vitis lubrusca*), are widely adapted to more than two-thirds of the country. They will grow just about everywhere except the deep South, but are most at home in the North, Northeast, and middle border states east of the Rocky Mountains. "California" grapes (*Vitis vinifera*) are European-type grapes originally grown in this country only in California because of their natural lack of hardiness and their adaptibility to a dry climate. Some of these make good wine, some are dessert grapes and some are raisin grapes.

In recent years, efforts to cross vinifera grapes with the American lubruscas have been quite successful; the new grapes are fairly hardy in the North and are capable of making wine which connoisseurs judge to be *almost* as good as European wine. (Almost but not quite, as there is a myth that no native American wine could possibly be as good as *imported* stuff.)

The third important branch of the grape family is the southern muscadine (*Vitis rotundifolia*). Muscadines are sweet, grow quite rankly if given a chance, do not make big clusters and have a fairly tough skin in comparison to other grapes. They make excellent jelly and juice and fairly good wine. Plant breeders are now crossing rotundifolias with viniferas and developing good wine and dessert grapes that will grow well in the South.

American slip-skin grapes, which I like to call the Concord family, are divided into three rather distinct types by color: blue-black, red and white, the last being often more yellowish or green than white.

A fellow by the name of E.W. Bull of Concord, Mass., gets credit for giving us the famous Concord—in 1852, which makes it one of the oldest fruit varieties still widely grown. Concord is the parent of other old favorites, Worden and Moore's Early. Fredonia is a pretty good blue-black which ripens very early compared to Concord. That makes it a good variety in northern-most areas where growing season is short. Beta, Bluebell, Buffalo, Early Giant, Sheridan (a very late variety), Steuben and Van Buren are other widely adapted blue-blacks. Carman and Extra are blues which originated in Texas and will tolerate southern heat better than most. Van Buren is considered by many grape lovers to be the best of the very early blue-blacks, but it gets mildew easily. Pierce is another warm climate grape of the Concord type and is grown, or used to be grown anyway, even in California. Buffalo has high dessert quality. Alwood is a good one for organic gardeners, being particularly resistant to diseases.

The grandpappy of the red American type grapes is the Delaware, first identified in 1849—before Concord became well-known. Though not quite as hardy as Concord, Delaware is still grown widely. Wyoming is an early red more suitable to the upper limits of the country than Delaware, though I don't know where it is still available. For those coldest climates, Red Amber, a newer variety developed in Minnesota, is available. It's very hardy and ripens in midseason. Catawba is a well known variety used mostly for wine in the Great Lakes area. Caco, a cross between Concord and Catawba, is light red in color, of good quality, but hardy only to mid-North, not upper North. Urbana and Vinered are late ripening reds, the former best for eating, the latter for wine. Yates is another late red which keeps well in storage—as do Urbana and Vinered.

The standard for the white labruscas is the Niagara, which has been around a long time too, and like Concord will grow just about everyplace except in an active volcano. Ontario and Portland are good northern whites that ripen earlier than

The grape is used to make raisins, pies and unfermented juice, but it has long been valued for another purpose. Noah, who was the first on record to have planted a vineyard, also was the first known to have experienced the grape's inebriating potential.

Niagara. Lake Emerald, light gold in color, was developed in Florida for that area. I don't know if it is still grown in the south, but if you live there and hanker for a labrusca grape, this would be one to try. To find one, keep your eyes open as you drive, walk or jog. Develop the habit of watching for gardens. When you see something that resembles what you are looking for, stop and talk to the gardener who owns it. Say something nice about the garden. Most gardeners love to show what they have to an appreciative audience. When your visit is over, you will probably leave with a promise to share any cuttings of grapes (or anything else) you want. Or better yet, you can talk a trade of cuttings.

Seedless Grapes

Another group of grapes are seedless varieties, most of them crosses between seedless viniferas and American labruscas: Concord Seedless, considered to be the product of a sport rather than a cross; Interlaken, Himrod, Lakemont and Romulus—all whites—plus Suffolk Red. I do not recommend these grapes to organic growers (unless you live in a very dry region) because all except the first are somewhat more susceptible to mildews than more standard varieties. (If you *do* live in a dry region of moderate climate there are other special seedless raisin grapes that might grow better for you.) There is no organic spray to control mildews as far as I know, unless you make an exception for sulfur as being "natural" sprays.

Most of the grapes in this group are small in size. With what little experience I have had trying to grow them, they seem quite weak in growth habit in my soil and did not taste nearly as good as plain old Concord, Niagara and Delaware. Your soil, climate or skill might be better than mine for these grapes, but my experience says that your time is better spent with other grapes.

I would make an exception if you are a great lover of grape pie. If so, Concord Seedless makes excellent pie. We have used regular Concords for pie but that is a job. You have to squeeze the pulp out of each grape, simmer the pulps, and put them through a collander to get rid of the seeds, then add the skins again. That's a lot of trouble for a grape pie.

There are a few dessert grapes that belong in a separate category. They resemble and are related to vinifera but are hardy over a wider area. They do not have the slip-skin characteristic of true American grapes, and are generally less hardy and more subject to mildew than Concord type grapes. But if you are successful in growing them, they make table grapes or wine of good quality. Century I is a blue grape in this category. Vines are vigorous and productive but fairly susceptible to both black rot and powdery mildew. Alden is a reddish black grape with meaty berries. It needs to be pruned short, and we'll talk about that later. A new grape, Lady Patricia, makes nice clusters of golden berries but is not very hardy above zone 4. Golden Muscat and the reddish black New York Muscat are both dessert grapes. The latter makes excellent wine.

Wine Grapes

In more recent years, especially with the increasing general interest in home winemaking, plant breeders have undertaken to develop many more fine wine grape varieties that can be grown in the North. This undertaking actually started in Europe, where the native American root aphids called phylloxera were unfortunately transmitted and caused so much damage to susceptible European grapes. The only way to fight the affliction, breeders found, was to graft viniferas onto resistant rootstock of American grapes.

With much of the phylloxera threat ended, plant scientists, especially in France, went one step further. They began crossing European varieties with American grapes, hoping to combine the excellent wine quality of their grapes with the hardiness and disease resistance of ours. Many of these "French hybrids," as they are generally called, have been tested in the United States, and the developmental work continued here, especially at the New York Experiment Station at Geneva, N.Y. The work is also going forward in Ohio, Pennsylvania and as far west as Missouri. Hopefully, there will arise many more fine wine-producing regions in America than we have now.

You as an individual gardener interested in wine or in selling wine grapes to home winemakers can now benefit from all this work. I don't know how widely the new hybrid varieties are available, but I see from my 1974 catalog from

Bountiful Ridge Nurseries, Princess Ann, Md., which just arrived, that a number of them are listed. The New York State Cooperative Fruit Testing Association catalog contains the largest selection I know of, but I have a hunch that Ohio and Pennsylvania nurseries have a good selection too. The breeders who have tested the grapes say they are suitable for American climates and will make wines similar in quality to French and German wines.

There are a great many of these varieties, but I'll list here only those that seem to be practical for home gardeners, especially organic gardeners. The names are just as they appear in the catalog: the vitaculturist's name followed by the common name. I have not grown any of these grapes myself.

Seibel 13053—Cascade. Very early, hardy, and mildew resistant.

Kuhlman 188–2—Foch. An early blue which makes a red Burgundy type wine. Hardy, resistant to downy mildew but only moderately resistant to powdery mildew.

Baco 1—Baco Noir. Blue-black, makes a Burgundy wine, good for heavier, poorly drained soil. Disease resistant.

Seibel 10878—Chelois. Should be pruned short. Fairly resistant to mildews.

Seibel 7053—Chancellor. Ripens about the same time Concord does. A very good winemaker, it is very susceptible to downy mildew early in the season. Seems to resist the disease later on, after leaves have hardened a little.

Seibel 5279—Aurora. A white winemaker (all the preceding make red wine), very early, good to eat out-of-hand too. One of the hardiest of the fine wine varieties. Resistant to downy mildew, but somewhat susceptible to powdery mildew.

Seyve-Villard 12–375—Villard Blanc. A late yellow grape. Not exceptionally hardy but does well from mid-north to middle states. Resistant to mildews.

Seyve-Villard 5–276—Seyval. Midseason grape. Must be pruned short. Resistant to most diseases.

If you want a fine American wine that is more practical for organic culture, try Missouri Riesling. This grape is native to middle America, having been discovered along the Tennessee-Kentucky border. It is hardy up to about the 43rd degree of latitude (Nebraska-South Dakota border) and is the most resistant *wine* grape to mildews. It makes a semi-dry wine.

Muscadines

Every organic gardener in the South—from Delaware to Texas—should grow muscadine grapes. They are among the most reliable small fruits because they are seldom seriously affected by diseases or insects. Muscadines rarely need pest control measures in the home garden.

Some muscadines are "perfect-flowered," which means they are self-fruitful and will pollinate other pistillate (female) varieties. In a muscadine vineyard, 1 of out every 9 plants should be perfect-flowered. Varieties that are "staminate" (or male only) also exist, but they bear no fruit. Nurseries always identify which varieties are self-fruitful and which require other self-fruiting varieties to pollinate them.

Scuppernong is the oldest muscadine—in fact the oldest grown variety of anything still available in the U.S., I think. Some plants still bearing fruit are well over 100 years old. Scuppernong is so old that the name is colloquially applied to any bronze-colored muscadine. This has led to the selling of "Scuppernong" plants of widely varying quality. Seedlings of Scuppernong are not the true variety; only cuttings from the original (which was found in the wild) merit the name, and good nurseries sell only that kind.

Popular muscadine varieties vary by area, and they produce fruit in different colors. Scuppernong (bronze, self-unfruitful), Thomas (reddish-black, self-unfruitful) and Hunt (black, self-unfruitful) are grown primarily in the Carolinas and Georgia.

Burgaw, Dearing and Magoon are the self-fruitful pollinating varieties best suited for the home garden. Burgaw is reddish-black, hardy, and vigorous at least as far north as North Carolina and probably as far north as Delaware, at least along the coast. Dearing is a late green grape with high sugar content. Magoon does better farther south than either Burgaw or Dearing and is used there as a pollinator.

Other popular varieties that are not self-fruitful include the Creek, a very good late grape, reddish-black, high in acid, good for the northern parts of the muscadine growing region; Dulcet, black, excellent vigor and productivity, ripens early —about Sept. 17; Higgins, a large white grape, grown frequently in the area immediately south of Atlanta. Topsail, grown mostly in the southern part of the muscadine belt because of limited hardiness, received the highest rating from USDA in quality. However, as is often the case with high quality fruit, Topsail does not always bear heavily. In USDA tests, it will bear 45 pounds of fruit where Hunt bears 101, or Thomas 74, or Scuppernong 63, or Creek 91, or Dulcet 87. But lower average yielding is not always a mark against a garden-grown variety where quality may be more often desired than quantity. Yuga is another variety often grown in the southern regions, but not recommended for the northern muscadine belt. It is reddish-bronze and late maturing.

Five new varieties, which USDA has not yet evaluated extensively, are all perfect-flowered types: Albemarle, Chowan, Magnolia, Pamlico and Roanoke—all developed in North Carolina; Bountiful, Chief and Southland from Mississippi; and Cowart from Georgia.

Starting With Grapes

Let's assume you have not grown grapes before but intend to try some. If you live in the northern two-thirds of the country, you should start your venture with an American labrusca variety in most cases. That means a Concord, or one of the newer blue-black varieties like Concord: Price, maybe Buffalo or, even better, Alwood if you intend to grow the grapes strictly organic. Next choose a white grape. Niagara is still the best in my book. You might try an Ontario too, as it ripens 2 weeks before Niagara. If birds are bad in your area, you should plant mostly white grapes because the birds will not bother them as much as the blues and reds. No, I don't know why. Green and white are just not colors that attract birds like reds and blues do.

Then you'll probably want a red (in a home garden, a couple of vines each of red, white and blue is most desirable, even if you are not all that patriotic). Make it Delaware, unless you live in Minnesota-type winters. Then try Red Amber. Where winter cold is not so much of a problem, Caco may be better than Delaware to fight mildews. Moored and Naples are worth betting on, and Yates is good if you want to try to store grapes fresh.

The plants will cost you about a dollar-and-a-half each, maybe a little more, maybe a little less. At the way inflation is sweeping along, by the time you read this, prices will no doubt be even higher. (You can remedy this by growing some cuttings from a neighbor's or friend's plants, which I'll discuss later.)

Plant your vines about 8 feet apart in rows about 9 feet apart. That would give you about 600 plants per acre, if you are interested in growing that many. Remember to think ahead, before you plant. Are you going to clean-cultivate the vineyard? If so, your rows should be wide enough to accommodate whatever mechanical cultivator you plan to use. If your vineyard is going to be small, mulched and hand-weeded, space between rows can be closer.

Make sure your planting hole is large enough to accommodate the roots, then set the vine in it so it will be neither any deeper nor shallower than it was in the nursery. You don't have to trim off broken roots but some gardeners do. Put good topsoil around the roots, but no fertilizer in the hole. Don't let roots dry out during the planting process as grapes are particularly vulnerable to drying. When you tamp the dirt around the vine, leave a slight depression right around the plant to hold water when it rains.

Let the vine grow willy-nilly the first year, on the ground. But by winter have posts set along the row about every 15 feet or so to hold 2 wires, one about 2 feet off the ground, the other 5½ feet off the ground, directly above the other. You may want a third wire, or use only 1, but the 2 wire Kniffin system will probably be the best training method in the home garden.

The Key to Pruning

I hasten to say that you don't have to train your grape vines in any particular way. There is no "right" way to prune, but rather there are a number of methods depending upon your personal purposes and preferences. The simplest "general" rule to follow is to remember that, overall, the varieties we're talking about (labruscas) will do okay if left with no more than 60 buds per mature vine, and no less than 40. What's a bud? Grape vines have joints or nodes between lengths of stem. Count every joint a bud.

The old rule in grape pruning is to leave 30 buds for the first pound of wood removed at pruning time, and an additional 10 buds for each subsequent pound removed. If you removed 5 pounds of wood, there would be about 70 buds left to produce the crop. That rule may be helpful to an expert pruner, but it leaves me in quiet desperation. How do you weigh a bunch of scraggily grape vines?

I think most efforts to tell a beginner *how* to prune a grape vine should be preceded by telling him *why* he's pruning it. When I finally grasped the "why," the how more or less took care of itself.

The first lesson to memorize is this: grapes develop on the *current* year's growth. From the buds you leave at pruning time, new growth will develop in spring, and on the new growth comes the fruit in summer.

Lesson number 2: The best fruiting new growth develops on *year-old* wood—on the branch vines that grew last year. They are more or less pencil-thick and smooth (older vines develop a fibrous bark on them).

Lesson number 3: Stop and think a bit how a grape vine grows, if left unpruned. The pencil-thick smooth vines of this year become the older, fibrous-barked vines of next year. The new wood that develops each year is necessarily farther away from the trunk of the grape plant. In 5 years, the good fruiting wood would be 30 or more feet away. Not only would that mean a completely unmanageable vine, but at that distance from the trunk, fruit usually will not develop in quantity or will not mature properly.

So the idea is to keep the grape plant producing within prescribed bounds. To do that effectively, you must *renew* the fruiting arms that you have trained to grow from the main trunk. So, you cut out the arm that grew the fruiting vines last fall, and you leave a new, year-old arm that is growing off the main trunk or off the base of the old arm near the main trunk. You also leave a spur—a vine cut back to 1 or 2 buds, near the base of the arm you are leaving this year. From that spur will grow the vine that will become the cane you will save for fruiting next year.

One more step: On the cane that you save for bearing this year's fruit, there will be branching vines at about every node. Cut these back to 1 or 2 buds. Some people leave 3.

By now you will think you have butchered your vines unmercifully. You may have less fruit than you would have if you had not pruned so severely, but you will have much higher quality fruit (and enough in my experience for all your needs), and you will have thinned out the vines enough to give good aeration, thereby giving some defense against blight diseases.

I follow the procedure I just described on 4 arms of each grape plant, trained to the 2-wire Kniffin system. You can also use a 3-wire (6-arm) system, or a 1-wire (2-arm) system. Twine is best for tying vines to the wire.

Sometimes, grapes are trained to overhead trellises or arbors, in which case the vines are allowed to arch overhead, tied to the framing, and maintained somewhat longer than in a vineyard. The gardener in this case prunes mainly to make the vines cover the arbor for shade, and the grape bunches to hang down under the arbor about head high. But the same renewal system is used to keep the vines productive.

Grape vines are also used to form a protective wall to keep a patio or yard private from the prying eyes of nosy neighbors. In such a case, pruning rules change. You are more interested in vegetative growth than in fruit. You will not renew as severely, but allow the vines to grow and branch in a thicker pattern.

It is very difficult to write a description of how to prune a grape vine. But just remember that you want to keep cutting out the old fibrous-barked canes, and you can't go wrong. You may even want to renew the main trunk of the plant every 10 years or so. All you have to do is allow a cane that comes up from the base of the plant to grow up alongside the old trunk. Then a year later, cut the old trunk off, and train arms from the new trunk to the trellis wires.

Some gardeners pinch back the fruiting vines after the grapes blossom and begin to form berries. The thinking behind summer pruning is that the vigor of the plant will be forced into making nice clusters of extra-high-quality berries. Experts disagree over the merits of this practice for fruit production. But tip pruning will encourage more branch

Perhaps the most popular approach to trellising grapevines in the home garden is the arbor. Fruiting arms are trained overhead. Come summer, the foliage will create shade, and the grapes will be easily accessible overhead.

growth and therefore make a thicker "wall" where privacy is desired.

Many types of vinifera grapes—and the new hybrid wine grapes—are pruned "short." That is, almost all vine growth is cut back during dormancy to a main trunk and a few short spurs. Since habit of growth among these varieties will vary according to soil and climate, you should find someone with experience in your locality and see how they prune. If that is not possible, you will just have to experiment. Try pruning some the way you would Concords. Prune at least 1 back to the main trunk. Compare yields.

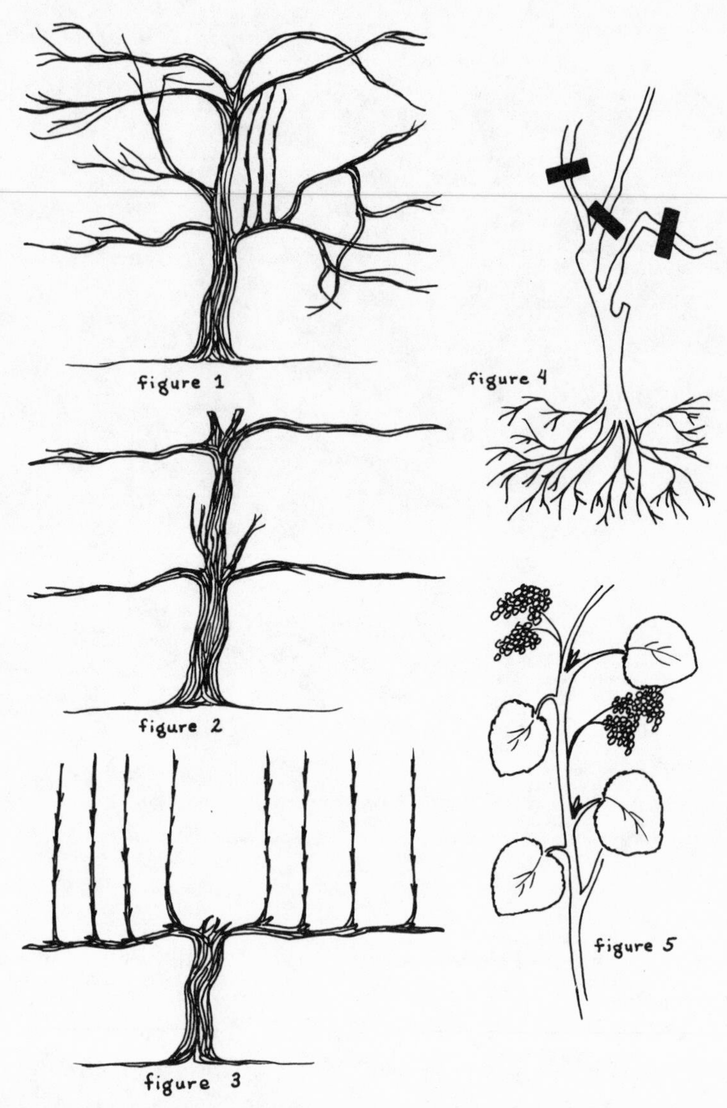

The grapevine (Figure 1) needs pruning to produce the best fruits. Figure 2 depicts the mature vine after it has been trained to the Kniffen system. The long arms are year-old canes; the short stubs will develop into the next season's bearing canes during the summer. The vine in Figure 3 has been pruned for the Upright system, with stubs left to produce the next season's bearing arms. Figure 4 shows how to prune a young plant before it is set in the vineyard. A flowering bearing arm is detailed in Figure 5. By limiting bud growth, one can control the production of each arm.

Pruning muscadines is about like pruning Concord types. However, the fruiting arms need not be renewed every year or every other year, as with Concords. Once established, fruiting arms are good for 3 or more years, until they begin to decline in productive vigor. A renewal cane is then allowed to take its place. Branching vines along the fruiting arms of muscadines will develop in excessive numbers. Cut some of them back to 3 bud spurs (1 about every foot along the arm) and cut the rest off completely.

All grape vines develop tendrils which will grow around the trellis wire, arbor or anything else within reach, to anchor the vine. On muscadines, these tendrils will wrap around the vines themselves, girdling and killing them. So don't let tendrils encircle the main trunk or fruiting arms of your plants. If a tendril now and then girdles a spur vine, that's not much of a problem since you will generally have more spur growth than you need anyway.

With muscadines you must be especially severe and firm in your pruning. Left untended, the vines develop into a dense, tangled mess. Should you inherit some old muscadines on that piece of property you intend to turn into your little paradise, proceed with stout heart in this manner: If some year-old canes are present, whack off all other growth, including all but 1 main trunk (there will undoubtedly be several trunks). If no year-old wood is there to make new fruiting arms—and there probably won't be—then cut off all branches, saving only 1 main trunk. Train new arms from that trunk the next year.

If the plant is hopelessly tangled and utterly impossible, cut off everything about a foot from the base of the plant. New canes will emerge during the next growing season. Select 1 for your trunk and cut off the rest. The following year, select the 4 vines most suitable for the 4 fruiting arms and cut off the rest exactly as you would go about training a Concord to the

Kniffin system. Cutting back to a 12-inch stump often gives you a more satisfactory "new" plant than cutting back to a main trunk, though the former means you will wait a year longer for fruit. If you have more than 1 old vine to restore, cut back only half of them that far, allowing the others to go on giving you what fruit they might until your restored vines come into production. Then whack back the others.

Prune all grapes while they are dormant. Most pruning is done in late winter or very early spring, so that any winter-injured wood can be cut out. But grapes I have tended (as far north as Michigan) never winterkill, so I have generally pruned anytime from December until March.

Many gardeners will not prune grapes in late spring when a cut vine will "bleed." Sap will ooze from a cut, and it does look as if the plant's vitality is being drained. Actually, plant experts say bleeding will not harm the plants anymore than tapping hurts maple trees. But gardeners like myself, who tend to assume without reason that plants have almost human qualities, don't like their canes to bleed like that, no matter what science says. I prune in February before pruned canes will bleed much.

The bleeding characteristic of grape vines is used to advantage by woodsmen. Lacking water from another source, an experienced outdoorsman will cut a wild grapevine about head high, then cut the piece loose at ground level. Water will drip from the lower cut, if the top if kept elevated. When the oozing quits, he will cut about a foot more off the top, causing still more sap to drain out the bottom. A hard way to get a drink, but better than no way at all. Wild grapevines will supply some liquid this way any time of the year.

Soil Fertility and Grapes

Since the ultimate aim of pruning grape vines is to keep vegetative growth in balance with fruit production, how much you will have to prune will depend to some extent upon the fertility of your soil. The poorer the soil, the fewer bunches of grapes you want to ripen because you want each plant to make only the fruit that will mature with good quality. If the soil is rich, you can allow more buds to grow to a larger crop because the fertility is there to support a bigger yield.

Grape roots grow down into the ground a considerable distance, which is probably why some wine grapes are said to do well on "poorer" soil. The deep roots find moisture even in dry weather, and they feed on minerals too deep to have leached away. Because of root depth, grapes may not respond to surface applications of fertilizer the way grass or vegetables will. Best to stick to a long-range program of nutrients slowly released from manures, mulches and rock phosphate. Certainly grapes in the home garden will never need a heavy dose of chemical nitrogen or triple superphosphate.

Some wine grape growers, even commercial ones, believe in a strict organic fertilizer program for full-flavored wine. Too much potassium from potent chemical fertilizers can reduce acidity in the wine, they say. Too much nitrogen can hamper proper ripening.

All grapes will produce better on poor land than most any other crop. New York farmers found that out years ago. When through improper farming methods or lack of sufficient fertilization, a field would no longer raise a decent corn crop, the farmers would plant it to grapes. However, grapes will do better on good soil.

On poorer soils, iron and/or magnesium deficiencies may show up. Magnesium-deficient leaves of white grapes turn yellow between the veins, which remain green. On blue grapes, the leaves develop purplish and reddish discoloration. Iron deficiency yellows new growth on vines. But as I pointed out in the chapter on blueberries, such discolorations may be symptoms of a number of diseases. If you have a reasonably good garden soil, suspect micronutrient deficiencies *last.* But if soil tests do show deficiencies, iron and magnesium (and other minerals) are available in chelated form from fertilizer companies. I'm not chemist enough to know exactly what chelated means, but usually chelates are organic forms of minerals which organic growers can use with a clear conscience.

Grapes in the North and West do not do so well if you try to grow them in sod or allow grass to gain a foothold around the plants. Clean cultivation is much better. Mulch is good too. Horse or cow manure bedding is fine. Poultry and rabbit manure are pretty rich for grapes. Seaweed and green alfalfa are good mulch for grapes. Think of the mulch on grapes in terms of its slow release of nutrients rather than moisture-saving ability. Grapes don't need moisture the way most other fruits do. Given a situation where I had only so much mulch and had to decide whether to put it on the grapes or on some other berry crop like raspberries, I would most certainly put it on the latter.

Green manure crops grown between the grape rows and plowed under will aid your organic fertility program immensely. Buckwheat and clover are good for this purpose, especially on heavier soils.

A soil management program for muscadines does not have to follow all the advice given for labruscas. For instance, these grapes are sometimes grown on a permanent sod cover, the sod kept mowed during summer, disked lightly early in the fall and seeded to a winter cover legume. More often though, the ground is cultivated between winter covers of crimson clover, vetch or peas.

In the muscadine regions, soils are often deficient in nutrients. Fertilizer—nitrogen, phosphorous and potash—will be more necessary. Magnesium may be in short supply and should be suspected if the muscadine leaves turn yellowish.

Feathered Pests

If your experience follows mine with Concord-type grapes in the East, your biggest problem will be birds. They peck a hole in each grape and suck out the juice without eating the pulp. But even as I say birds are the chief menace, I am aware of a mystery. Last fall, for reasons I cannot discern, the birds did not bother my grapes—the first year in 9 that they did not. Sometimes I think the creatures of the garden divine what I'm about to say about them and then act contradictorily just to remind me that man knows only a little about nature. But that is what makes gardening so interesting.

The best protection against bird damage is to slip a paper bag over each cluster of grapes when they are still small and green, and close the end of the sack with a rubber band. Believe it or not, this was standard practice in vineyards years ago when labor was cheaper. My father-in-law, who raised grapes for commercial market, says that once you get the knack of it, bagging grapes this way is not as painfully slow as it sounds.

I'd certainly recommend the practice on a small arbor. I must confess that I don't do it. Despite birds we get enough grapes to eat, and the ones our feathered friends peck can still be used for jelly. But we rarely have any big bunches of purple grapes nice enough for a table centerpiece or for selling.

Another way to beat the birds is by growing more of the white grapes, as I have mentioned. Birds just don't bother them the way they do the purples, blues and reds.

Plastic nets or cheesecloth draped over the arbor may keep out a few birds. My experiences with these materials have not been very satisfactory—though better with grapes than with blueberries. Unlike blueberries, the grapes hang down, underneath the vines, and the netting does not lie so close to the fruit that birds can peck the fruit right through it.

Other Grape Menaces

The second worst menace of grapes in the Northeast is the Japanese beetle. He loves grape leaves. We have never had so many as to keep us from getting grapes, but the voracious bugs can weaken the plants by eating so many leaves. We arm ourselves with jars, and in the evening when the beetles are slow of movement, we brush them off the leaves into the jars. My wife and I have at times gathered a jarful each in less than half an hour. Bug traps are available from garden stores that will catch thousands of the beetles every day. You can also purchase Milky Spore disease, much publicized in organic and natural gardening magazines. It can keep beetle populations down to a level you can live with, used as directed.

Japanese beetles constitute another of my little mysteries of life. They can occur in large numbers 1 year and few the next without any apparent change in control measures; they can heavily infest a specific area and be scarce a few miles away. I've observed this in the surrounding neighborhoods just north of Philadelphia and can find only 1 explanation—and that one only tentative. Where there are fairly large areas of sod, under which the beetle grubs like to overwinter, the green iridescent bug seems to be much worse. So when you see starlings waddle their peculiar way across your lawn, pecking into the brown grass in late fall or early spring, be thankful. They are after the grubs.

You shouldn't really blame Japanese beetles for eating grape leaves. Grape leaves are very tasty, as anyone brought up on Greek cooking knows. In June, when the leaves are still succulent and tender, pick some. Prepare them with rice, ground beef, onions—just as you would stuffed cabbage.

The most serious diseases of grapes in the Midwest and

East are black rot and mildews. The former shows itself first as dark black, sooty spots on the grapes when they are still green. The grapes, instead of ripening, shrivel up like raisins. Sometimes the rot doesn't progress beyond the skin spots, leaving the grapes marred but usable. A copper spray will control black rot if it is not too bad, but organicists will just have to hope the weather turns drier. In excessively wet summers, no chemical will do a very good control job on fungal diseases. My Concords and Delawares seem to be pretty resistant to black rot, even in the wet summers we have had in '72 and '73. But I have some yellow grapes, gift of a neighbor, variety unknown, that are quite susceptible.

What is true of black rot is equally true of those grape nemeses, downy mildew and powdery mildew. Some of the newer fungicides give some control, as will sulfur, but an organicist can only prune well for good aeration, grow his grapes in full sun and bank on resistant varieties. I have no problem with my Concords.

Grapes are extremely susceptible to herbicides. Drift from weed sprays can harm grapes hundreds of feet from point of application, if the wind is right. If you cannot stop road crews from spraying along the road next to your property, grow your grapes as far away as possible from the road, upwind from prevailing breezes.

The larvae of the grape berry moth eat grapes. The worms are greenish and about ⅜ inch long. A worm will eat out 1 grape, then move into another contiguous to the first. When you pick the grape cluster you find the 2 grapes stuck together. These worms are seldom numerous enough to be a real menace. They start feeding on the blossoms, and so right before bloom, organic growers can get some measure of control with a dusting of rotenone.

Brown aphids and leaf hoppers do some small damage but not enough to concern you. Root aphids, the phylloxera mentioned earlier, have been controlled by resistant rootstocks. If

Trained to a wire trellis, the grapevine exposes its fruits for easy harvesting. The fact that the grapes hang below the foliage when so trellised makes it possible to protect them from birds by draping the vines with cheesecloth or plastic netting.

you buy fancy wine grapes grafted to resistant rootstocks, make sure when you plant them that the graft union is an inch or 2 above ground level so that roots do not develop above the graft. You'll be able to tell the graft union by the slight bulge at the base of the plant. By the same token, any new shoots that come up below the graft should be pruned out immediately. The grapes that would develop from the rootstock would be inferior—possibly wild American grapes.

Maybe I shouldn't call wild grapes "inferior." Actually the wild fox grapes all over the northern U.S. make excellent jelly. And a lot of us farming people used to think wild grape wine was great when made by a Minnesota farm woman and served by Minnesota farm girls whose eyes sparkled like the wine in our glasses. And the blossoms of wild grapes! The almost perfect perfume. (Nothing is perfect.)

Muscadines are the cultivated fruit least bothered by insect and disease. Well . . . *among* the least bothered anyway. Black rot, bitter rot and leaf spot may become troublesome, but rarely in the home garden.

The most noticeable loss from bitter rot usually occurs right before maturity, when a few grapes may decay and shatter.

Leaf spot (cercospora) affects leaves, not fruit. Small brown spots on the leaves induce a yellowing, and if the disease gets serious, many leaves will fall. But that seldom happens, and even if some leaves do fall, the plant is not particularly harmed.

The grape berry moth, the grape flea beetle and the grape curculio all occur in muscadine vineyards but are of little consequence except in large commercial plantings.

Propagating the Vineyard

Once you have a grape vine growing well, it is no trouble starting new plants from it. A length of vine growing close to the ground can be rooted by layering. Just bury a portion of the vine an inch or 2 below the soil surface, leaving most of it, especially out towards the end, above ground. After it roots, cut off the vine with its new roots, and transplant.

Cuttings will root fairly well too. I take cuttings from year-old wood—from the vines I prune off in February or December. Cuttings should be about a foot long—this is my own formula, not quite what others recommend—and containing 3 bud nodes, 1 on each end and 1 in the middle. I thrust the cutting into the ground until only the top bud node sticks above ground. That leaves 2 nodes below for roots to grow from. I have a thin steel rod I use to poke a hole in the ground, into which I then shove the cutting. I set the cutting in the ground immediately after I take it from the mother plant. Anytime in early winter or early spring when the ground is not frozen is okay.

A grape cutting must be planted right side up—the cut end toward the base of the plant goes into the ground; the cut end toward the end of the vine above ground. If you are taking a dozen or so cuttings at once, as you probably will, you may find when you start to stick them in the ground, you literally can't tell which end is up. I've had it happen to me.

To prevent that kind of stupid situation, I make the bottom cut, right below the bottom bud node, *on a slant;* the top cut, right above the top bud node, *square across.* Then I know which end goes into the ground—and the slanted bottom cut goes in easier too.

When warm weather comes, leaves will appear from the top bud node. With a little luck and adequate rain roots will be forming underground, and the cutting will "take hold." If it doesn't rain, keep the ground moist and mulched around the cutting. Just let the vine ramble on the ground the first growing season. If you have put the cutting right where you want the vine to grow permanently, you won't have to transplant it.

I've set cuttings in both December and February–March, and in both cases, I've gotten about a 50% survival, without going through the rigamarole of burying bundles of cuttings upside down under sand until they "callous" as the experts advise.

It takes about 5 years for grapes to come into full production—both labruscas and muscadines. Yields vary greatly, especially with muscadines, but 4 tons to the acre is common, and 5 tons definitely in range of any grower. On a per-plant basis, 35 pounds of labruscas is a good average to shoot for, but it's possible to produce more than that. Muscadines, per plant, may yield up to 100 pounds, and you should figure on an average of 50 pounds.

Pick grapes only when they are completely ripe. They will not ripen further after harvesting. Grapes keep pretty well even when ripe. Use a sharp knife to cut bunches. Be careful when picking. If they can find an entrance, wasps and bees can get inside a grape. Approaching from a "blind" side, you cannot always see the rascal eating out the inside of the grape. In the act of picking, you may get stung. Usually though, the wasps and bees drinking on that sweet juice are in much too good a mood to sting unless you squeeze them threateningly.

Concord-type grapes can be stored over winter or at least for a couple of months. Maybe muscadines and viniferas can be too, but about them I don't know. Store only choice, perfect clusters of labruscas, *fully ripened on the vine.* Pack in layers of dry, clean sawdust or maple leaves in boxes kept in a cool, frost-proof location.

Of course the best way to store grapes is to make wines, jellies and juice. And there are plenty of books now to tell you how. Good eating.

Chapter Nine

The Cranberry and Its Wilder Cousins

We were having sorghum pancakes for supper, the flour ground in our blender out of our own cane sorghum seed—the stalks of the same plant providing us with the molasses for the pancakes. But alas, my wife informed me that as far as sorghum molasses was concerned, our cupboard was as bare as Mother Hubbard's. And two months yet to go before we could boil down some new maple syrup. Would we have to resort to plain, store-bought syrup to sweeten our gourmet-special sorghum pancakes?

Fortunately, we could do better than that. We just happened to have some lingonberry jam, which my wife, Carol, was willing to use to make up some heavenly lingonberry syrup. But the pancakes were hot and I was in no mood to wait. I just spread the jam on and ate. Scandinavians know a thing or two about good eating—they and their descendants in this country have been praising the lingonberry for centuries.

The fruit is best described as an upland cranberry or a mountain cranberry. It looks like a small cranberry—it *is* a small cranberry—but does not have to have a bog to grow in.

You can usually find it wherever cultural conditions are right for wild blueberries, with which cranberries are closely related.

The lingonberry is also called the cowberry (I don't have the slightest idea why), the red whortleberry (whortleberry being an old name for blueberry) and foxberry. Its scientific name is *Vaccinium vitis-idaea*, which is quite a mouthful but probably safer to rely on than foxberry. I imagine there are about 10 different kinds of berries that are known in 1 locality or another as the foxberry.

In the Northwest there's another wild cranberry *(V. quadripetalum)*, which the Indians taught white settlers to use.

The Wild Cranberries

But none of those should be confused with the more "famous" eastern wild cranberry, which haunts the bogs of north-central and northeastern states. "Haunts" is the right word too, I think. There is something a bit mysterious, even ghostly about cranberry bogs. All my life I have been hunting for a real, honest-to-goodness wild cranberry bog, with no lack of grizzled old farmers urging me on. Their bogs were always of the variety: "Back-yonder-in-the-woods-someplace-thet-grandaddy-told-me-about." I don't know how many such ephemeral cranberry bogs I have searched for in Minnesota, Wisconsin and New Jersey, without much luck.

Writers on the subject have been of little help except to lend more mystery to my search. Nelson Coon in his interesting and helpful book, *Using Wayside Plants*, teases with titillating vagueness: "I discovered a fine growth of them [cranberries] 1 fall within 20 feet of a very heavily travelled highway in the backwaters of a famed New England river." Notice he does not say which heavily travelled highway nor what famed New England river.

My sister Teresa has fallen under the spell of the folklore

that surrounds wild cranberry bogs. In a letter telling me about the small farm she and her husband purchased, she wrote: "Somewhere back in the woods, say the oldtimers in these parts, there was once a cranberry bog, but we haven't found it yet." Eventually I will join them in that search, but I know in my bones already that the bog will have vanished on the air of oral tradition where it has existed for so many years. But I will not give up hope. Next to wanting a never-failing spring and a clear-running brook, what I'd treasure thirdly is a real-for-sure bog with wild cranberries in it.

Actually, no one should be too disappointed or surprised not finding cranberry bogs where oral history says they once existed. Without doubt, drainage of farm land has destroyed bogs that did exist a hundred years or so ago. Farmers, sometimes out of greed, sometimes on the advice of "experts," but most often in dire need of cash to pay their bills and debts, have plowed up every inch of land a tractor could be driven on without sinking or falling off. Then they drained the sinkable land. (If the government would subsidize bulldozers, they'd soon level hillsides now too steep to farm and turn the country into one unbroken crop field, all with encouragement of the efficiency experts and the sellers of "money-saving" technology who rake off what little profit the farmer makes from mining the hell out of the soil.)

Both the wild cranberry and the lingonberry are low, creeping plants very hard to spot except in late fall when the berries are sparkling red, or in the case of bog cranberries, in early spring when a thick stand will almost glow with a purplish hue. The lingonberry, with its shiny green leaves, would make a beautiful ground cover if it could be established in a thick stand. I don't know of anyone who has tried that yet. The first requisite would be acid soil—pH of around 4.5. But since the lingonberry will survive winters in zone 2, where it can get cold enough to freeze your eyeballs if you

The cranberry is native to America and was used extensively by the Indians. It was first cultivated about 100 years ago on Cape Cod.

look into the wind too long, an attempt to domesticate the lingonberry in this manner might be worth the effort. Ask any wise old farmer of Swedish extraction living in Minnesota. The trouble with lingonberries, at least the fault of the ones I have found, is that each plant or vine produces only a few berries. It takes quite a large patch of them to yield enough berries to make your picking worthwhile.

The Domestic Cranberry

The domesticated cranberry, *V. macrocarpon*, is, of course, an important commercial crop in Massachusetts, New Jersey, Wisconsin, Oregon and Washington. The necessity of flooding the crop part of the time makes the cranberry impractical in the home garden, but this does not necessarily have to stop adventuresome gardeners. The flooding of commercial cranberry bogs is not a requisite for growing the crop, but is mostly a protection against both excessively cold and excessively hot temperatures and a control for certain pests. On the West Coast, for instance, cranberries have been raised successfully with sprinkler irrigation instead of flooding, believe it or not. If cranberries mean that much to you, you *could* grow a few if your climate is not too cold or too hot. You will also need acid soil—a pH of around 5 being just right. Keep the plants well watered during the growing season and mulch them heavily with peat during the winter. The biggest problem with raising cranberries this way is weed control, but you could hand-weed a small patch.

Certainly the cranberry is worth the trouble to grow it, if you want yours free of chemicals. Not only is the berry a good source of vitamin C, but it is believed to help clear up bladder infections. Doctors don't know why. But if you are hospitalized with bladder trouble or suffer infection in connection with childbirth, you'll probably find cranberry juice as part of your diet. It may be an old bit of folk medicine but researchers have some evidence that it works.

Cranberries will not keep well fresh. One of the commercial grower's biggest problems is to control various rots in the harvested fruit. But the berry freezes well.

"False blossom" is about the only serious disease of cranberries. It's a virus disorder that in the late '20s nearly wiped out commercial production. As the name indicates, the flower pedicels do not develop correctly, being upright rather than drooping normally. The petals then develop abnormally—short, wide and splotched with both red and green color.

The virus is carried by a leaf hopper. Over the years, USDA scientists have succeeded in developing varieties, like Beckwith, Stevens and Wilcox, that the leaf hopper doesn't find so tasty. No leaf hoppers, no virus. Secondly, flooding the bogs shortly after the leaf hoppers hatch in late May and early June helps control the bug. Roguing out diseased plants is not practical except on very small plantings. But at any rate, the various control measures now in force have successfully reduced the danger of false blossom.

The Highbush Cranberry

The "highbush cranberry" is not a cranberry at all, but since the fruit looks something like a cranberry and can be made into jelly and sauce using the cranberry recipes, I might as well include it here. The highbush cranberry is a member of the viburnum family, technically *Viburnum americanum* or *Viburnum trilobum*. It develops into a bush about 10 feet high. The fruit starts ripening in late July in zone 4, a little later as you move north, and persists through the winter unless the birds eat it. Indeed, providing winter bird feed is the main advantage of the shrub—that and the cheerful red color that it adds to a drab winter landscape. The bush is definitely a northerner, extremely hardy from zone 4 to an Eskimo's eyebrow. *Viburnum americanum* will flourish on the upper Great Plains, an area which has never shown much hospitality for berries.

Periodically a nursery here or there gets all worked up over the highbush cranberry in an effort to sell a bundle to gardeners. At that point, *V. americanum* becomes a delicious jellymaker for the northern climates. There's nothing dishonest in that claim, the definition of delicious being somewhat relative. Delicious compared to what? The berries will jell easily, being high in pectin, but you will have to sweeten them plenty if your taste is anything like mine.

I might add as a caution that another viburnum, *V. Opulus*, a native of Europe, is sometimes grown in cities in this country because of its resistance to smoke pollution. It looks a good bit like *V. americanum*, but the berries are too bitter to eat, even jelled.

To repeat, the greatest value of the highbush cranberry is to feed birds and as an ornamental in the North. That value is not to be sniffed at, as I will try to show in the next chapter while discussing other offbeat berries.

Chapter Ten

Unusual Ornamental Berries and Native Fruit

One of those brown, drab, soft-cold days of early December had cast its melancholy spell on me. I walked listlessly past my lifeless gardens looking for some reminder of the vigor of summer past, some promise that growth would come again. Even old Gray-wing, our resident mockingbird who permits no other feathered fellow near his dominion among the multiflora roses, seemed despondent. He sat on the tip-top of the spruce near his lair, for once quiet. The future held only snow and ice; he knew it as well as I did.

As I ambled somewhat aimlessly beyond the gardens, I became aware of a faint, high-pitched "screeeee," not unlike the sound an angry shrew makes—though I doubt that's a helpful description. The noise had been in the air, I'm sure, all the while, but I was only now conscious of it. It seemed to come from everywhere and yet nowhere.

Suddenly, the trees along the edge of the woods came alive with sleek, grey birds about the size of robins. Their high-pitched murmurings had such a ventriloquistic character to them that it took some concentration to determine that the sound actually was emanating from the birds. They ignored me totally, sweeping in flurries down upon a small wild haw tree laden with miniature yellow apples.

Cedar waxwings, I immediately deduced. I had seen some earlier in summer, but not in such numbers. I slipped back down to the house on the run, grabbed the binoculars and raced back to the woods.

Focusing in on the waxwings, I was most pleasantly surprised to find the birds were not cedar waxwings, but Bohemian waxwings, comparatively rare birds, which I had never seen before. The rich brown-red under tail coverts were the dead giveaway.

For over 2 hours, I watched the birds feast upon that wild fruit. Even though they jerked the fruit from the twigs in the most ravenous manner, swallowing each whole in a single gluttonous gulp, the waxwings maintained that elegant, princely air that has always charmed birdwatchers. It was a show that I would have paid more to see than any theatrical performance, yet it cost me nothing except a sore neck. And it turned a dull December Sunday into one of the nicest days of my life.

Finally, the flock, as if by some mysteriously communicated consensus of opinion, burst into the air and flew away, not to return. I went up to the tree and tasted one of the yellow berries. Bitter, but not too much so. And for a bird, a luscious dessert in December's niggardly larder, no doubt. I remembered that once in early spring I had contemplated cutting down the thorny little tree as worthless brush. How glad I was now that I had changed my mind out of mere whim for the blossoms. Now I got the axe and cleared away all the other brush and small trees crowding in on this lure of Bohemian waxwings. This is one tree that will stand as long as it is in my power to protect it.

The moral of this minor birdwatching drama is that the more varied one's garden plants, the more interesting one's gardening life, because a variety of plants attracts a variety of birds and a variety of birds means something more than better insect control.

Successful organic gardening demands as complete and complex a biological chain as can possibly be maintained. Orchestrating that chain so that no link becomes aggressively dominant over another link is the real work of a gardener. Nature abhors a vacuum; she abhors even more sameness, uniformity, specialization. Show me gardens of exquisitely trimmed hedges, neat rows immaculately kept, lawns as smooth and shorn as a nylon rug, and I will show you a man doing violence to nature.

Only man can grow vast fields of a single species of plant, and he can do it only at great cost in chemical and mechanical control of weeds and disease. In nature a pure stand of plants of almost any variety is inevitably struck down. If the bug that infests them in turn becomes too numerous, then, like the gypsy moth, it too will decline. Overpopulation starves deer; lemmings rush to their doom. Right now, man himself is the menace, his numbers dominating the earth in some areas to the extinction of other parts of the biological wheel of life.

A Variety of Berries for a Variety of Purposes

In his own small way, the true gardener bows to this fact and does his best to insure maximum variety. In berry gardening, he will not only grow the berries already discussed, or as many of them as fit his climate, but he will cultivate judiciously many varieties of wild and ornamental plants that produce fruit for birds, animals and, possibly, human use too.

These berries are not honored primarily for their tastiness, but they are often attractive enough to birds to keep them from your choicer berries. Or, more importantly I think, these berries attract birds that eat insects too: plants like juniper, bayberry, hackberry, honeysuckle, snowberry, thornapple, haw, holly, wild cherry, pokeberry, barberry, viburnum, multiflora rose, mountain ash, dogwood, bittersweet, serviceberry, huckleberry, sour gum, wild grapes, cotoneaster, sumac, nannyberry, winterberry, cedar and autumn olive. From my own gardens and from technical garden lists, I've composed a list of the fruit-producing plants you should consider for your lawns and gardens.

——1. The dogwood family: *Cornus.* A person generally thinks of dogwoods as lovely blooming ornamentals of spring. But the blossoms develop berries much loved by fall migrating birds, and the red fruit is as beautiful as any flower. (Incidentally, dogwood used for firewood gives more heat than even hickory. I don't know many people who would burn one, but I have.) We are fortunate in having many; they like our naturally acid soil.

There are several species of dogwood besides the usual kind. *Cornus alternifolia*, blue dogwood, also called pigeonberry, grows as a small shrub or small tree to 15 feet high and has bluish fruit. *C. canadensis*, the bunchberry or crackerberry, is a low, woody herb with greenish flowers and scarlet berries. It ranges across northern North America. *C. rugosa* is the green osier of the East, with blue fruit.

——2. The bearberry was called by the Indians kinnikinick; by scientists, *Arctostaphylos uva-ursi*. I prefer bearberry. It's a trailing evergreen vine with scarlet berries relished by wild animals as well as birds. It's a great ground cover for acid soil in the Northeast if you can get it started. Folklore says a tea brewed from dried bearberry leaves is good for urinary troubles.

——3. The blue berries on juniper and some cedar trees are extremely high in vitamin C. Birds evidently know that too. They were regarded as an excellent preventive for scurvy back when that disease was prevalent.

——4. *Gaylussacia baccata* is the true huckleberry, a delight for both man and bird. It will grow in partial shade too, which gives it an added advantage as an ornamental. Huckleberries, strictly speaking, do not belong to the blueberry family though the 2 fruits look and taste a lot alike. A huckleberry has 10 comparatively large hard seeds, while blueberry seeds are soft and small. *G. frondosa*, a similar species, is sometimes called dangleberry or tangleberry. *G. dumosa* is called gopherberry. Both are edible but not very tasty to humans.

——5. The garden huckleberry, sometimes called—in a burst of advertising fervor—the wonderberry, is an altogether different fellow. Luther Burbank developed it and the less meticulous vendors of plants can call it "huckleberry" conveniently only because it does have blue-black berries. Actually the prolific berry-maker is a member of the *Solanum* family closely related to the tomato. If you like green tomato pie, then you will like garden huckleberry pie too. Add a little lemon juice to zip up the flavor a bit. It's not good to eat out of hand.

Some seed and plant catalogs carry an item called "Saskatoon blueberry." I'm not sure what it is, but it's not a blueberry. It's often called "juneberry" but that name is ap-

The garden huckleberry does not provide as good eating as the blueberry and it belongs to an entirely different family.

plied to many berries. When in doubt, always look for the scientific Latin name. If not given, be circumspect about the plant being advertised.

——6. The *Myrica* family contains 2 excellent plants for feeding birds and of possible use for you. *Myrica cerifera* is the wax myrtle, a tall shrub or small tree with grey berries. *M. pennsylvanica,* the regular bayberry, is a smaller plant, growing 3 to about 8 feet tall. Both are common on the East Coast.

There's no reason why you couldn't make bayberry candles like our ancestors did, or at least make enough bayberry wax to scent tallow wax candles. Pick the berries around November 1 and boil them in water until the mixture reaches the consistency of thick syrup. Strain out the berry skins and seeds. When cooled the wax hardens, of course. Reheat and melt it for dipping candles.

Last summer I watched members of the Goschenhoppen Society of Montgomery County, Pa., who are dedicated to preserving the old ways, dip homemade bayberry candles without using candle molds. It's very simple really. A piece of wick is dipped into the hot wax and drawn out immediately. Some wax adheres to the wick and dries. Then the wick is dipped and dried again. Eventually the wax builds up with succeeding layers to a candle of normal size.

Bayberry wax can also be used to make soap, using it in place of animal fat tallow.

The bog myrtle *(M. gale)* is closely related to the bayberry. The fruit is more like a nut and makes an aromatic spice.

Bayberries prefer acid soil. Where blueberries or azaleas do well, you could grow them for ornament. Bayberries will draw many birds, especially the myrtle warbler.

——7. Wintergreen is another commendable ornamental with many uses. Scientifically, that's *Gaultheria procumbens*, but you may also hear it referred to as checkerberry, partridgeberry, boxberry, mountain tea and other colloquialisms. The showy scarlet berries and leaves are good breath fresheners. The leaves make tea from which the genuine oil of wintergreen can be distilled. The plant grows on the East Coast from Maine to Georgia and inland approximately as far as Pennsylvania. It's evergreen, as the name indicates, with white flowers. Though hard to transplant from the wild, it can be propagated from cuttings of half-ripened wood. It will grow only in very acid soil.

Moxieberry or creeping snowberry is a whitish fruit of the *Gaultheria* family. The fruit has an aromatic birchy flavor. Moxie is the old eastern colloquialism for soft drink. Perhaps a drink like root beer can be made from the berry, for all I know.

A more well-known member of the *Gaultheria* clan is *G. shallon*, called salal on the Pacific Coast. It's an evergreen shrub with reddish to purple-black berries about a half-inch in diameter. Salal berries are nothing much for eating out of hand. They are plentiful and kids will eat them. They make passable jelly. West Coasters use salal branches for floral greens. The tough, evergreen leaves last a long time after cutting and so make good Christmas decoration.

——8. Autumn olive, *Elaeagnus augustifolia*, is one of the very best plants for wildlife cover and feed and makes a fairly good ornamental hedge. Wildlife experts almost always recommend it along farm fence rows and around ponds to attract wildlife. The bush develops yellowish egg-shaped berries with a kind of silvery scale on them. Leaves are silvery underneath too. Another species, *E. commutata*, is even more silvery and is in fact called silverberry. Both are very hardy.

Sixty years ago, another species of the same family, *E. longipes*, was being touted as a possible "new" commercial fruit called Goumi. Listen to this description: "A graceful and handsome bush 5 or 6 feet high bearing a profusion of silver-white leaves and most abundant crops of cinnabar-red and gold-flecked berries. Whether considered for ornament or for fruit, it is one of the best of the many excellent shrubs which have come to us from Japan." Now who could resist that? Such prose has done little to further horticulture (who has Goumi today?), but what an exciting influence it has had in the development of advertising literature.

——9. The soapberry, of the *Sapindus* family, grows in Florida. The Indians learned that the pulp of its yellow to orange-brown fruit lathered easily.

——10. Birds will only eat snowberries (*Symphoricarpes albus*) when they are very hungry. But I grow the bush anyway, because its pure white, marble-sized berries all winter long are nice to look at. They fit in beautifully with a bouquet of holly at Christmas time too. A more important advantage of snowberry: bees love the flowers.

In the same family is coralberry or Indian currant, *S. orbiculatus.* It has white flowers, reddish-purple berries and crimson foliage in the fall.

S. occidentalis is what some people are refering to when they say wolfberry. But wolfberry is a label for other wild species too. This particular bush has white fruit and is very hardy.

——11. Hackberry or sugarberry *(Celtis)* will lure birds to your garden. It's a fairly large tree with a range from central to southern U.S. *Celtis laevigata* has orange to purple-black berries, one-third of an inch in diameter.

——12. Pokeweed or pokeberry (*Phytolacca americana*) makes new growth each spring from the roots, like asparagus. In fact the tender young new shoots taste like asparagus—they are even preferred to asparagus by some wild food collectors. I hate to throw a wet blanket on that claim, but young pokeweed is better by comparison only when it is very fresh and the asparagus it is being compared to is not. All things being equal, asparagus is better than poke, believe me. If you do boil the young poke shoots, boil in 2 waters. Throw the first water away.

Birds eat the pokeberries in the late fall when the latter are fully ripe. I am always amazed by this because the seeds are supposed to be poisonous—though the fruit itself is not. I am not about to try that out on my own body. The juice makes a good dye, good food coloring and passable ink, according to botanists. The fleshy roots of pokeweed are poisonous too, and so are the leaves and stems when they turn red late in the year.

You really should not be alarmed that people eat plants, parts of which are poisonous. Rhubarb leaves are poisonous too. So are wild cherry leaves. There's poison in apricot seeds. The thing is, in many cases where a plant or part of a plant is labelled "poisonous" (it makes a crop of articles for Sunday supplements and gardening magazines every year, as editors are real suckers for this subject), the "poison" is not all that deadly unless taken in huge quantities in refined form. For instance, there is enough nicotine, more than enough, in a pack of cigarettes to kill you if you drank it all at once—which is certainly an excellent argument for not smoking. However, one pack of the awful things is not going to kill you even if you inhale every puff.

I make this point so that you use your head around plants, and so that you don't become overly alarmed. As an example, here is a passage from a gardening book: "Such a green potato, as well as any green shoots which may arise, is rated as a deadly poison, not to be eaten cooked or green." I'm certainly not going to deny that, but I'm not going to agree with it wholeheartedly either. Potatoes that are exposed to the sun a little will turn greenish on the side the sun hits. We cut the green skin off but eat the potato. Never hurt me yet.

Castor beans are poisonous too. We used to put them down mole holes. Folklore said the moles would either go away or eat them and die—I was never quite clear on that. Anyway, castor oil which comes from the bean was forced down my throat when I was a kid. The oil was a "medicine," the bean a "poison." I'm not hungry enough to test all these claims out, but do you blame me for doubting? And as far as I am concerned, if taste is any measure, castor oil ought to be put on the top of the poison list. Do any poor children have to take that awful stuff today?

——13. The juneberry or serviceberry, *Amelanchier canadensis,* also called shadbush or sarvistree, is one of the best trees for bird food. You will also do yourself a favor by planting one: the fruit makes jam, jelly and pie. The tree, which can grow 50 feet high or more, is also an excellent ornamental. The white blossoms please you in spring, the red berries ripen to purple in summer and the foliage changes to yellow in the fall. The fruit contains more vitamin C than citrus and has no serious bug or blight problems.

——14. The salmonberry, a sort of low grade raspberry, is a native of the Northwest. Both yellow and red varieties exist, but the taste is not much, or so I'm given to understand—I've never eaten any. Birds and children will eat them, but rather half-heartedly.

——15. In the East, the orange-red raspberry-like wineberry grows wild—an original transplant from the orient which long ago escaped from gardens. Gardeners allowed the escape

The pale red wineberry is slightly sticky to the touch. It does not have as full a flavor as the raspberry.

because the wineberry never gained the favor the raspberry has. However, the wineberry isn't all that bad, for a wild berry. We gather them enthusiastically. They are a little seedier than red raspberries, a little juicier and more sour. Wineberry canes are a rusty orange in color and hairy, which easily distinguishes them from black raspberry and blackberry canes often growing near them.

——16. The buffalo berry (*Shepherdia argentea*) should be better known than it is, especially to gardeners in the northernmost parts of the country. The plant is available commercially—Gurney Seed and Nursery Co., Yankton, S.D., sells it. Some people know it as wild oleaster or silverleaf. Seventy years ago, botanists were discussing the commercial possibilities of buffalo berry optimistically, but the berry never could compete with our regular fruits. It makes good jelly and good winter feeding for the birds. It is also an excellent ornamental with its silvery leaves and red berries. It will make a hedge in dry, cold and windswept locations where nothing much else will—a perfect plant for the northern plains. *Sheperdia canadensis* is even hardier and occurs naturally farther north than *S. argentea*, but the berries are not as good.

——17. Hansen bush cherry (*Prunus besseyi*), sometimes called sand cherry, is available commercially from quite a few nurseries. It is not really a cherry, but is a fairly good fruit for pies and excellent to lure birds away from your choicer fruits. The plant is also very hardy.

——18. I mentioned *Viburnum trilobum*, the highbush cranberry, in the previous chapter. It is one of the most important berries for birds you can grow. Another species, *V. lentago*, has bluish-black fruit and may grow up to 30 feet tall though it usually gets only half that height. It's called nannyberry or sheepberry too.

——19. The winterberry (*Ilex verticillata*) is the best berry in the holly family for birds. Often called black alder, the winterberry is probably the prettiest of all red berries that endure through most of the winter—or until the birds eat them. It is often used for Christmas decoration in the East and is even cut and sold for that purpose. Its range is the East Coast as far south as Georgia.

Ilex glabra is another interesting member of the holly family. It grows wild in bogs—but will grow in gardens in acid soil. Its main importance, for the organic gardener seeking to build a total biological chain in his garden, is that it is a good bee plant.

Smooth winterberry (*Ilex laevigata*) is very similar to *Ilex verticillata*, but the berries are more orange than red in color. If you hear someone talking about hoopwood, this is the plant.

——20. Sumac (*Rhus glabra*) is a very useful plant and one of the most beautiful. Where it grows well, it really does grow well, and because it becomes so common in those areas, a lot of people look at it as a weed. That's unfortunate because sumac keeps birds from going hungry—and much more. Just this morning, I cut some sumac branch sections to use as spiles for the annual tapping of maple trees. Sumac wood has a soft center (the smaller shoots) which can be quickly reamed out to form tubes. I usually split 6-inch lengths of branch, scrape out the soft core to form small "troughs" rather than tubes. The tube trough becomes the spile I insert in the tap hole, and the maple sap follows the groove out and drips in the sap bucket.

Sumac seeds can be crushed or bruised in water to make a drink that tastes about like pink lemonade. Add sugar to taste. Be sure to strain the "lemonade" through a cloth before adding sugar and drinking to get out the fine hairlike particles that will be present. The seeds used to brew the beverage should be collected during a rainless period, as their flavor can be washed off.

Some people make pin money gathering sumac leaves and selling them to the tanning and dyeing industry. Sumac leaves are rich in tannin. You should contact a Department of Agriculture office nearest you for information on how to get the USDA bulletin on collecting sumac leaves. I understand that users of tannin in the U.S. (including the beer industry) can not get enough of it anymore from their traditional sources in Japan. The idea of growing sumac commercially, long talked about, may become a reality.

There is poison sumac, but it is easily distinguished from the good kind. The poison stuff has white berries in loose drooping clusters. The good kind has red berries that stick straight up in the air in large, tight clusters. Poison sumac will treat you just like poison ivy does—they're first cousins.

Variety Beyond Berries

Not all birds eat berries, and even those that do don't eat them all the time. Birds want a varied diet, and you want them to have a varied diet. So in addition to berries, I try to provide a variety of other foods for the birds, bees and wild animals.

Oak trees are number 1 on my list, mainly because of the acorns. Together with other nuts (beech, walnut), acorns form what wildlife people call "mast." Mast is the principal food of squirrels and wild turkeys and a supplemental food for some birds. Acorns are also excellent food for hogs.

In spring during the migration season, the warblers heading north in their bright new plumage favor the pin-oaks because at this time the little green measuring worms are hanging from the trees, suspended on their glistening web strands, spinning their way to the earth. The warblers eat them and in the meantime provide me with hours of the most exciting kind of birdwatching—warbler watching. The chipping sparrows, who stay for the whole summer and usually hatch out 2 or 3 young ones, eat the worms too. I have watched mother chipping sparrow gather as many as 7 of the green worms in her beak at once and fly back to her nest with them.

Back and forth all day she works. How many worms is that?

In late summer, grackles and bluejays take over the upper regions of the oaks. They quarrel with the squirrels over the small acorns and peck at the pieces the squirrels inadvertently drop.

Weed and grass seeds draw birds too. That's why I don't mow my lawn frequently. Goldfinches and ground sparrows like dandelion seeds especially. The patches of grain I grow in my gardens—wheat, rye, barley, buckwheat and oats—lure more than starlings and pheasants. I left a few stalks of cane sorghum for the birds last fall, and to my delight a towhee, a bird that rarely comes to the garden, pecked at the seedheads all of 1 afternoon.

Another bird attraction with which we are blessed is a type of cherry tree, growing wild in the woodlot behind us, which I have not seen anywhere else. It is not a wild black cherry nor a chokecherry, but a type of black sweet cherry. My theory is that it is a seedling from a sweet cherry out of an old garden. This land was cultivated a century ago and then allowed to revert to woodland. Most likely there was an orchard where that woodlot stands now.

The cherries are small, mostly seed, but very good. More importantly, the fruit does not suffer the depredations of brown rot like my domestic sweet cherries do. I am now propagating the "wild" trees on my property so I can substitute them for the "better" sweet cherries.

Wild Habitat

What the birds love most about my gardens is the "hedge" that surrounds the property. The all but impenetrable border is composed principally of multiflora rose with a mingling (and I use that word advisedly) of Japanese honeysuckle and bittersweet. Having confessed that, allow me to add hastily that I recommend multiflora rose, honeysuckle and bittersweet only to stout-hearted and foolhardy gardeners. No timid soul, be he of the peace-loving, pensive nature generally associated with gardens, or be she the delicately winsome type also thusly associated, should be forced to face the savagery of vigorously growing multiflora, honeysuckle or bittersweet hedges.

Each winter I go forth to do battle with that hedge, for if I do not slash and cut it back to some semblance of hedge shape, it will within a very few years spread across my whole garden and that of my neighbors, conquering all other growth in its march. If you wish to war upon other nations, you need neither guns nor bombs. Merely plant multiflora rose along their borders.

I exaggerate of course, but only a little. And I must also defend multiflora hedges (as one who grows to love an honest adversary after years of competition). There is no cheaper wall of privacy around a piece of property, nor anything as hard to get through. My hedge has on occasion even turned back dogs and small boys, thus earning my undying gratitude. The blossoms in June are pretty and fragrant, though the honeysuckle is far more aromatic—the only justification for *its* existence. In the fall and winter, the red hips from the rose blossoms add beauty, especially against a background of snow. The hips are healthful food for man and bird. The bittersweet, by January or February, also keeps birds from going hungry.

The birds like the hedge as much for protective cover as for food. Mockingbirds, robins, brown thrashers, cardinals and catbirds—and perhaps others that I do not know about yet—nest among the thorns. Hummingbirds suck nectar from the honeysuckle; chickadees and finches eat the honeysuckle berries.

But I must trim that hedge. At least every other year I cut it back to just a little higher than my head, and about 4 feet wide. I do half of the hedge 1 year, the other half the next. The more you prune multiflora, the thicker it gets, giving secure privacy within your "living walls." If you feel like working in your garden in a scanty swim suit, the only one who will notice is that brassy old mockingbird. And if your neighbors appreciate privacy, they won't complain if an occasional multiflora vine reaches out on their side of the hedge and snags them once in a while too.

Birdlovers should also grow a few pine and spruce trees. Evergreens make excellent winter cover for birds and provide nesting spots for many species. The seeds from the pine cones feed chickadees, nuthatches, and warblers.

A pair of mountain ash trees will bring flickers, waxwings and finches to your door. Pyracantha (firebush) will provide food in a pinch, though the birds will eat everything else first in my experience. Grosbeaks like maple seeds. And grow a few sunflowers and leave them in the garden unharvested. Most of the seed-eating birds will thank you. Remember too that undesirable English sparrows cannot open sunflower seeds, so they will not chase away other birds, as they are wont to do at your bird feeders.

Should you provide bird food? If you manage a large garden the way I have described, and your neighbors do likewise, and there is brush and woodland around, you don't need to feed birds except during periods of heavy snow or ice. In fact, feeding birds at other times means they will eat fewer insects and weed seeds in my opinion—ornithologists don't agree on this matter.

But do provide water if a natural source isn't handy. Put shallow containers around your grounds and change the water in them frequently. Not only will the birds be grateful, but the bees and wasps will be too.

All this will not rid your garden completely of insects, nor will birds stop eating your best fruit. But you will have fewer critical insect problems and the birds will steal far fewer berries than they would otherwise.

Chapter Eleven

Make a Little Money From Berries

I have alluded occasionally to the possibility of profits from selling berries. I have done so hesitantly, because I do not want to give you the impression, as some books do, that you can make money easily with berries, or that you can make a living from the berry business without much experience.

Many commercial berry growers do make a living—some of them a very fine living. But if you don't start small and learn as you move along, you will suffer many heartaches and most of them will be financial ones. And in the process you will lose the fabric of an ordered organic life: serenity, independence, the kind of deliberate freedom that urges us joyously to the garden every morning and evening of our lives.

But you can make *some* money from berries without investing (risking) a lot of money in the project—a rare situation today. You can get started, learn, make spending money and then reach a decision about expanding into more commercial berry production. All this you can do in a sane manner *before* you sell your soul to a bank.

A gardener's or farmer's first experience in selling berries usually comes when he has a surplus crop. He doesn't plan it that way; it just happens. Weather may have been exceptionally good, or the grower may have gained the skill to grow a bigger crop that he calculated he could grow. At any rate, he sells some. It's almost as easy as giving away dollar bills. People *like* berries. They may gripe at the price, but they buy anyway. Then they call back a few days later and want more. The capitalist lurking in the heart of every homesteader starts thinking.

Or maybe a friend asks the gardener how to start a berry patch, and would you, kind sir, tell what varieties I should plant and where I might get them. The gardener, being a practical soul underneath it all, suggests that his friend take some of his own excess plants—the suckers in the raspberry patch have to be pulled anyway and there's always too many runner plants in the strawberry bed. Then the friend generously offers to pay for the plants. The capitalist stirs again.

Finally, your sister is visiting you, trading gossip in the kitchen and raving about your elderberry jam. Much to your chagrin, she is already on her third slice of bread and jelly, the latter spread nearly a quarter-inch deep on the former. As if to justify her gluttony, she allows as how she sure would pay plenty if she could buy jelly that good in the store. The capitalist inside is now fully aroused.

At any rate, all 3 of these methods of marketing berries—selling plants, selling fresh fruit, and selling processed berries—can be developed into worthwhile sidelines. The amount of money you make will be directly proportional to the amount of work you do.

The First Decision: Labor

At the start of any berry project undertaken for profit, I think the first decision is whether to do it with or without outside hired labor. Since there is a physical limit to how many berries a person can pick in an hour, you can't plant more berries than you and your family can harvest unless you plan on using hired help. With hired help (assuming you can get it just when and only when you want it) you can plant a big crop and try to make a killing on it. Without hired help, you have to plant a little of each kind of berry so as to spread the harvesting over the entire summer.

The latter way has some merit. You can start selling quite early in the year, even in the North. As soon as the thaw is out of the ground, you can sell plants of raspberry, blackberry, blueberry—any berry that grows as a cane or bush. As soon as strawberries start showing signs of life again, you can lift and sell runner plants for resetting. Actually, strawberries can be transplanted any time during the growing season that you have enough moisture. After the setting-out season has passed, the berry harvest begins: strawberries, raspberries, blueberries, gooseberries, blackberries, fall red raspberries. Then it's back to selling plants again in the late fall, except strawberry plants in the North. Meanwhile, you have been making jams and jellies from excess berries, and through the winter, you can sell them. Check with local health officials to see what approval you need to sell processed fruit.

How many berries can a family conveniently handle—without becoming slaves? A good strawberry picker should be able to fill a quart box every 5 minutes in a patch that has not already been picked over. He might be able to keep up that pace for up to 6 hours a day or about 70 quarts a day—but that's working *hard!* If you sell them at 60¢ a quart, you make $42.00 a day. If you and your wife together pick 100 quarts a day, that's $60—for about 25 days, the normal length of heavy strawberry harvest, or $1,500 gross. Raspberries average out about the same in harvesting profits, because even though you can't pick them as fast as strawberries (they're smaller), they sell higher. A pint of raspberries may sell even higher than a quart of strawberries. The same is true of blueberries. Blackberries and dewberries pick as fast as strawberries. With cane- and bush-type berries, you won't tire as quickly as when picking strawberries—you don't have to bend and stoop as much.

One person with family help could handle ⅓ acre of producing strawberries, ⅓ acre of new strawberry plants for the next year's production, ⅓ acre each of blueberries, raspberries and blackberries. For fall red raspberries, ½ acre shouldn't be too much because you can prune them mechanically the way I explained earlier and because the fall-bearing varieties do not produce as heavily as the summer-bearing kinds. Meanwhile you can probably handle a few currant, gooseberry and elderberry bushes on the side, if you have a market for them. You will now be very busy. But once you learn how to spread the work load out over the entire year, as I have explained, you could handle it if your wife likes to help. If you have developed a good retail market, you'll have no trouble clearing a minimum of $4,000. Depending on your yields and the demand of the marketplace, you could make more.

If you hire harvest help, your best bet will be women in the neighborhood who can work odd hours, when needed. Teenagers during summer vacation are the next best alternative. Girls work better than boys. One boy works better than 2 together. Put the guys and dolls together and everyone will have fun, but not many berries will get picked during the partying.

Don't expect hired pickers to work as fast as you do. They aren't motivated that much and there's not much you can do about that. Therefore, it is much better to pay by the box than by the hour, if you can. Find out what the going rate is in your area. A good picker ought to make from ¼ to ⅓ of whatever the market price of the berries, in my opinion. If a box of strawberries is selling for 60¢, the picker should receive from

15¢ to 20¢ per quart box. But rates vary by locality, and you have to make your arrangements to satisfy your help.

Labor has always been the biggest problem for fruit growers. That's why they turn to machinery whenever possible. Even mechanical raspberry harvesters are beginning to look practical out in Oregon. But mechanical berry pickers are very expensive and their use compels commercial growers to expand to bigger acreages to pay for the machines. The gardeners this book is meant for don't want to turn pro—at least not that pro.

One solution to the labor problem in the berry business is the Pick-Your-Own or U-Pick system, where customers harvest your berries and pay for them at a lower price. Most U-Pick operations I know about have been successful for the owner and popular with the customer. People like to come out of their city environs and work in the sunshine once in a while. But you must develop the patience of Job and maintain an ever watchful eye. People who have not had experience in berry fields and orchards need a little education, and you have to give it without hurting their feelings. Insist that they pick clean as they go, not jump around from row to row wherever they see a big berry. Insist that children stay close to parents and behave themselves—or provide a playground for the kids some distance from the fields. Most important of all, you will need ample, handy parking space.

With strawberries, if not all berries, it often pays to take the first heavy picking or two yourself with perhaps the aid of a couple of veteran pickers. This is the time the biggest berries are ripe and fast pickers can fill a lot of baskets in a hurry. Afterwards, when only medium-sized and small berries remain, open the rows to U-Pick. Urban customers enjoy the picking and don't have to break any speed records to make a wage. Children can frolic among the berries without doing nearly the damage they could when the patch was producing heavily. And the later berries are almost always sweeter. So everybody wins.

Produce or Product

Jams and jellies are an excellent way to use up any berries you don't get sold or that have gotten *a little* overripe before you could get them harvested. But don't try to sneak bad berries that are beginning to rot into the jam. If you want someone to go out of their way to buy your homemade jellies instead of the easily available jars at the supermarket, you must maintain highest quality and taste.

Freezing berries to sell in the off season has not been tried by many homesteaders as far as I know. The cost of the comparatively large cold storage capacity needed stops them. However, someone is going to get serious about this idea someday and make it work for a small business. It certainly is much easier to process berries by freezing them than by making jams and jellies out of them.

I once helped build a walk-in freezer for a community of some 60 people who were trying to produce all their own milk, meat, fruits and vegetables. A room about 12 feet by 15 feet in a cellar was walled off from the rest of the area, triple-insulated with thick layers of cork panelling. The freezer unit came from a food locker that was going out of business. The cost of maintaining a freezer room of that size was about what it would cost to operate 4 regular freezers, I was told. But they got lots more room than they would have with 4 freezers.

Another possible money-maker is raising grapes to sell to home winemakers. Winemaking is very popular right now, as you know, and many people who like to make wine don't like to grow grapes. They are paying plenty for the fruit concentrate from wine hobby shops, and some of them could be paying plenty for your grapes instead. I have 2 winemakers ask me every year if I have extra grapes. Opportunity is knocking.

Of course, if you start thinking very long about wine, you'll be tempted to try to make and sell *that.* In Pennsylvania, a progressive law now allows small winemakers to sell their own wine retail, and quite a number of people are taking advantage of it. Pennsylvanians can get detailed information from their State Department of Agriculture at Harrisburg. If the Appalachian states had made similar laws regarding whiskey before the industry became dominated by a few powerfully rich companies, good bourbon would be cheaper today, I bet, and Appalachia would not be sunk in poverty.

The Marketplace

However you sell your berries, you have to have an outlet that fits your supply. If you are only going to harvest 100 extra quarts of strawberries, you can't expect to build a profitable roadside stand trade. Nor can you expect a big supermarket to be interested in buying your 100 quarts. The buyer at the supermarket may love your succulent, organic berries, but he must have a steady supply in order to do business. If he forsakes his regular suppliers and distributors to sell your berries for a week or 2, his suppliers may just forsake him when he needs them again.

On the other hand, small local groceries that do not carry a very big inventory of fruit will often welcome your excess berries. Be sure your quality is the highest and be sure to quote a price that allows the grocer to make a little on the deal too.

If you can make yourself famous in the local area for producing some specialty better than anyone else produces it, you can sell almost anywhere—in big supermarkets and small corner groceries. Let's say that you learn how to produce a very sweet golden raspberry that you started from a plant an

The small farmer or homesteader wishing to market his surplus crops and the consumer looking for fresh produce often meet at the roadside stand. Fresh berries can be a drawing card.

old farmer in Pennsylvania Dutch country gave you long ago. (This is not a farfetched example. An old farmer *did* offer me a golden raspberry plant several years ago.) Hardly anyone grows yellow raspberries, let alone good, big, juicy ones, so before long, everybody is talking about Joe Smith, who raises those golden raspberries, and how good they are. Perhaps the local paper sends a reporter out and does a feature on you. You now have got it made. You can sell golden raspberries anywhere in town easier than selling funny papers on Sunday. Give your product a label: Smitty's Golden Nectar Raspberry or something like that. Fifteen years of hard work and you can retire. All marketing studies show that there's more money in a labelled product sold wholesale over a wide area than in a whole bunch of nondescript products sold retail at one store. And, incidentally, you'll be able to sell your golden raspberry *plants* at premium prices too. Never be afraid of selling plants to start another gardener out. Any time you have a neighbor who admires the produce you sell and wants to raise some too, help him. Nine times out of 10 he isn't going to keep it up. But he will get used to eating good, garden-fresh food and when his own berry patch goes to pot, or when he gets tired of taking care of it, he'll be back to buy more berries from you than he ever did before.

Not many people reach such high estate of developing a famous brand name in the berry market. Most gardeners-turned-capitalists must sell retail to make any money. There are 3 ways to go with a small, retail, part-time business: 1) a roadside stand; 2) a customer route; 3) a personalized pick-up service.

The roadside stand (about which much has been written, some of which is even true) is a reasonably versatile marketing tool. You can operate a large one or a small one, full-time or part-time. True, if you aren't open every day, you are going to lose some business. But if you're after only a part-time income, you can't handle a lot of business anyway. Besides, if you've got something good to sell, local people will buy it, even if you're only open a couple days a week—*as long as they know when you will be open.* In other words, keep a consistent schedule. If you say you are going to be open Thursday, Friday and Saturday, stick to it.

I've been to roadside stands off the beaten path that can afford to be open all the time because there's no one there to wait on you. You buy what you want and put your money in a can. I hope—I wish—all of us lived where people were that trustworthy. However, if the stand is close enough to the house, you can keep a watchful eye on it and maybe make an honor system work for you.

At other stands, there's a bell rigged up which the customer rings for service. That way, you can get some work done in the house or nearby garden or barn while tending the store at the same time.

Don't let the fuel shortage deter you from starting a roadside stand. People may not drive out to a stand as often during fuel shortages, but veteran roadside marketers tell me that customers buy more when they do come. That's the experience of those who went through gas rationing during World War II too.

Many market gardeners of the past established lucrative routes over which they delivered produce to customers on a regular weekly schedule. Some marketers still do. The advantage is that you do not have people coming to your place all the time as with a roadside stand. A route is better if you like privacy around your house. Also if you already have a truck, or need one, it may be cheaper for you to go into the route business rather than build a roadside stand. I don't think I would do too well with a route; I talk too much. I'd never finish the day's deliveries. By day's end, I'd still be arguing politics with the third or fourth customer I'd met!

The personalized service approach is the least structured of all selling methods. It is not very efficient and is not tremendously profitable, percentage-wise or any other wise. But you don't have to work at it much either. What you do is sell whatever you have available to a very select group of people, usually close acquaintances, preferably well-heeled. You don't have to advertise; you don't even have to call them usually. They know. They look at the calendar and they think: "Say, ole Gene's red raspberries are ripe about now.

Guess we'll happen by." And so they come, and so I gladly sell the berries at a much higher price than I can get anywhere else. It all works out well because I have an Understanding with such acquaintances. They know that I would rather give them the berries. I know that they know that I would. But they know that I know that they know that I know they would be embarrassed to tears to barge into the driveway because Gene's got berries he wants to get rid of. By paying handsomely for them, we all preserve our pride and good humor. It's one of those games people play. And my son, who more than likely picked the berries and will get the money, is overjoyed.

Another friend and casual customer for our extra berries is a family that faithfully buys eggs from us. They drive over every week, and when the berries are ripe I sell them some. Usually I give them some too or sell cheap. After all, if they drive over here for eggs, what with gasoline being the way it is, that's worth some strawberries.

And that reminds me of the most successful way to "sell" of all for a part timer: giving. To give someone a box of strawberries I have labored long and hard to bring to fruition organically—that is a very special gift. Those special human beings who accept such a gift while understanding and appreciating it the same way as I do in giving it—well, that is what berry gardening is *really* all about!

Index